LAKE ERIE TECHNICAL WRECK DIVING GUIDE

LAKE ERIE TECHNICAL WRECK DIVING GUIDE

Erik A Petkovic Sr

ISBN 978-1-909455-30-6 (Hardback)
ISBN 978-1-909455-27-6 (Paperback)
ISBN 978-1-909455-28-3 (EPUB ebook)
ISBN 978-1-909455-29-0 (PDF ebook)

Cataloguing-In-Publication Data A catalogue record for this book can be obtained from the British Library.

Copyright © 2019 Erik Petkovic. All intellectual property and associated rights are hereby asserted and reserved by the author in full compliance with UK, European and international law. No part of this book may be copied, reproduced, stored in any retrieval system or transmitted in any form or by any means, including in hard copy or via the internet, without the prior written permission of the publishers to whom all such rights have been assigned worldwide.

All photos and illustrations are copyright Erik Petkovic, except where stated otherwise or out of copyright.

Cover Design © 2019 Dived Up from an original idea by Paul Lenharr. Front cover underwater photo of the wreck *John J Boland Jr* by Becky Schott. Back cover photo of *Barge F* by Vlada Dekina.

Printed by Lightning Source.

Published 2019 by

Dived Up Publications
Oxford • United Kingdom
Email info@divedup.com
Web DivedUp.com

Erik A Petkovic

Contents

Dedication   7
About the author   8
Using this book   9
Local diving notes   10
Glossary   12
Map   14
Acknowledgments   16

Introduction 17
1   Acme 21
2   Andrew B 28
3   Atlantic 30
4   Barge F 52
5   Cracker 55
6   Dunkirk Schooner 58
7   George J Whelan 73
8   John J Boland Jr 90
9   Junction 20 100
10 Mast Hoop 102
11 Oneida 103
12 Oxford 108
13 Persian 115
14 Saint James 122
15 Sir C T van Straubenzee 131
16 Smith 143
17 Stern Castle 149
18 Swallow 150
19 T-8 157

Suggested reading   159
Contributors   163
Index   164

For all those early wreck diving and technical diving
pioneers who inspired me to reach the depths:

Gary Gentile, Steve Gatto, Tom Packer, Gene Peterson,
Brian Skerry, Mike DeCamp, Billy Deans, Peter Hess,
Brad Sheard, Bart Malone, John Dudas, Harold Moyers,
John Moyer, Pat Rooney, Jim Bunch, Pat Clyne, Hank Keatts,
Bill Palmer, Cris Kohl, Garry Kozak.

About the author

Erik A Petkovic Sr is a technical wreck diver and maritime historian specializing in shipwreck research. He has been published in multiple international dive magazines and is the author of *Shipwrecks of Lake Erie: Volume One* (2017). A highly sought after speaker, Erik is available for presentations and seminars. He currently works in Washington DC and resides in Southern Maryland with his wife and two sons.

Using this book

> Divers conduct dives entirely at their own risk. This book is intended to increase their knowledge before attempting these sites, but it is no substitute for proper training, experience, preparation, organisation, support and supervision.

With the above warning in mind, the ratings system used in this book is intended purely as a quick reference guide. There are two wrecks which can theoretically be dived shallower than 130 feet (40 m) and which therefore receive the one star 'Deep' rating. General advice and most training agencies suggest that because of its ability to reduce narcosis the use of trimix is a very good idea below 130 feet. However, as air can still technically be used, and because it depends on divers' own assessments, only dives deeper than 185 feet (56 m) are designated 'Trimix' (and three stars) in this book. Everything in-between is rated 'Tech' (two stars). This is displayed for each dive as follows:

As always in deeper diving, the decision about how to conduct these dives is yours alone.

Local diving notes

Charters

There are not many charters which go to the deepest wrecks in Lake Erie. Most are smaller boats which accommodate few divers and visit the shallower dive sites in the Eastern Basin. Osprey Charters based in Westfield, New York, regularly take divers to all the wrecks presented in this book. They are professional, accommodating, and have operated their charter business for over 20 years aboard the *Southwind* — a 50 foot vessel capable of carrying 20 divers. They are extremely knowledgeable on all the wrecks.

Osprey Charters
ospreycharters.cochrane@gmail.com
www.OspreyDive.com

Accommodation

Motels

There are several motels within a few miles of Osprey Charters. I recommend contacting the following:

Holiday Motel
223 North Portage Street
Westfield, NY 14787
716-326-3741

Theatre Motel
7592 East Main Road
Westfield, NY 14787
716-326-2161

Campgrounds

There is a KOA Campground located nearby which, in addition to camping, also has some cabins available to rent.

Westfield Lake Erie KOA
8001 East Lake Road
Westfield, NY 14787
716-326-3573

Bed & Breakfast

There are many B&Bs located within a short distance of Westfield. These are all in the Lake Chautauqua region which is known for its wine, boating, and fishing.

Glossary

Bark — a sailing vessel with more than three masts all of which are square-rigged with the exception of the aft mast, which is fore and aft rigged.
Belaying pin — a long piece of wood narrow on one end, bulky on the other to which rope is tied in order to secure sails.
Bitt — a post used to secure mooring lines.
Bowsprit — a spar which extends forward of the bow stem typically above a figurehead or other ornamental scroll head.
Bow stem — the curved wood at the forward most point of the vessel where the hull plates come together.
Braces — 1. as a sailing term, the tackle used to change the sailing angle, 2. referring to the ship's structural support.
Canaler — a type of ship in the Great Lakes. Built small and narrow to maneuver the tight, windy canals and locks.
Capstan — a mechanical device used to raise an anchor, an older version of a windlass.
Cofferdam — a watertight structure built below the waterline which can be pumped out to perform construction, maintenance or salvage.
Crosstree — wood or metal beams which stretch perpendicular to the mast. Used to secure and support shrouds.
Cutwater — extra wood timber or metal plate used to strengthen the stem.
Deadeye — a block of wood with three holes for rope which is used to tighten sailing rigging.
Fantail — a vessel's rounded, overhanging part at the stern.
Fife rail — a band constructed around a mast that holds belaying pins.
Figurehead — a decorative and ornamental carving at the bow beneath the bowsprit.
Gunwales — the upper sides of a ship.
Hawsepipe — the means by which an anchor chain passes through a ship's hull.
Hogging arch — wooden braces stretched along a ship's deck to add support, e.g. hogging arched package freighter.

Propeller — with the advent of steam as a mode of propulsion, 'propeller' was a type of ship designation used to identify a steamship or propeller driven ship during the end of the age of sail.
Samson post — a strong bitt or post at the bow or stern of the vessel.
Schooner — a sailing vessel with two or more masts, all fore and aft rigged.
Schooner barge — a schooner which has had some of its masts removed and been turned into a barge. It would be towed by other vessels.
Scow — wide, flat-bottom vessel with square ends.
Spar — a catch-all word for a wooden pole, mast, yard, bowsprit, etc.
Texas — a structure or section of a steamer which includes the crew's quarters.
Tiercin — an Old English unit of measurement equivalent to 14 pounds still used around the Great Lakes in the early age of sail.
Tiller — a long piece of wood used as an early way to steer a vessel prior to the advent of the helm.
Transom — flat surface at the stern of the vessel.
Turnbuckle — a device used to tighten or loosen the sail.
Turtle — an upside down/overturned ship or wreck.
Walking beam — an early type of steam engine where a pivoted horizontal beam would work the piston.
Windlass — device used to raise an anchor.
Yawl — a small boat on a ship for escape, commonly referred to today as a lifeboat.

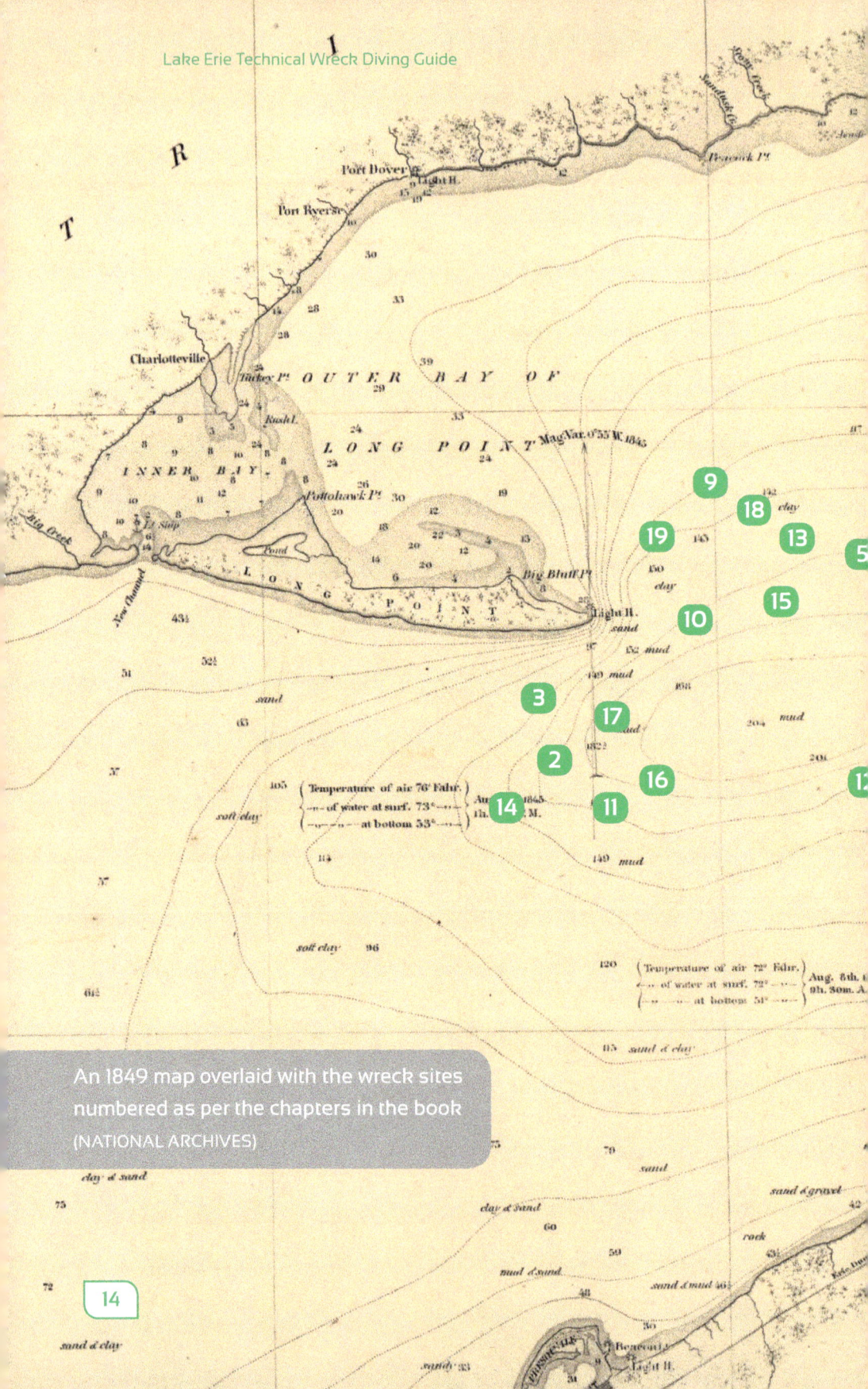

An 1849 map overlaid with the wreck sites numbered as per the chapters in the book (NATIONAL ARCHIVES)

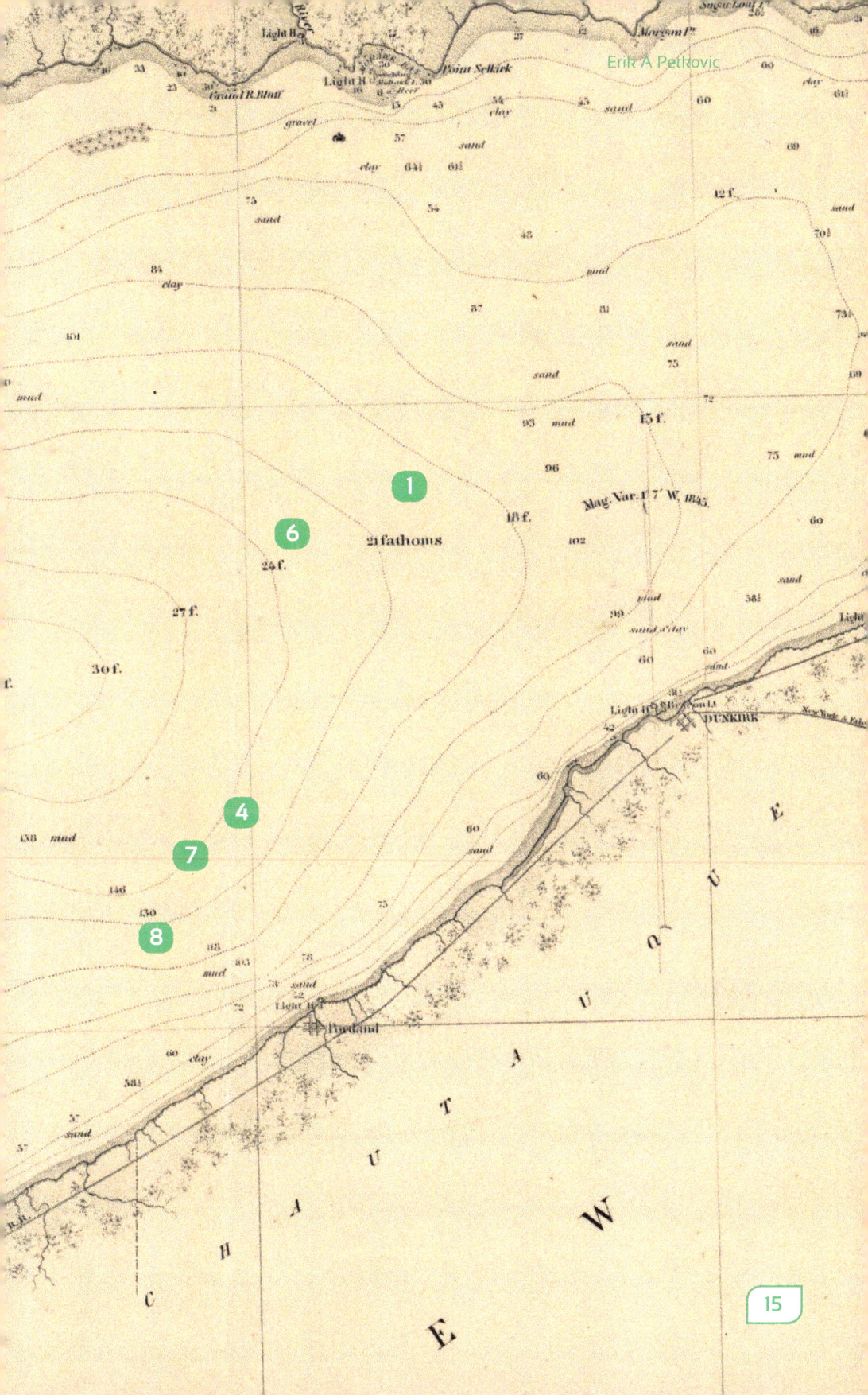

Acknowledgments

I would like to thank the following people without whose support this project would not have been possible:

My beautiful wife, Danielle, for her undying support in all my endeavors — from my ridiculous travel schedule for my real job to the countless hours researching, to the late nights and early mornings typing, reading and editing. You should have seen her face when I told her I was writing another book. Priceless.

My two sons, Joshua and Erik, Jr — future divers and history buffs. Thanks for always letting me use the computer (except during homework time). Remember, you can do anything you set your mind to. Anything.

I would be remiss not to mention all the gentlemen in the dedication who inspired me to challenge myself and explore sunken ships. Your passion and dedication to the sport is forever engraved in those who came after — including me. My parents always instilled the value of paying respects to those who came before. This one's for you.

All the phenomenal underwater photographers who provided exceptional images for use in this book. I cannot do with a camera what they can. They are the very best at what they do. They do not set the bar, they are the bar: Gary Gentile, Vlada Dekina, Cris Kohl, Becky Schott, Steve Gatto, Tom Wilson and Warren Lo.

Pat Clyne for opening your files on the *Dunkirk Schooner*. Here's to Peter Hess.

Garry Kozak for your side scan images. Nobody does it better. Thank you.

Paul Lenharr for the exceptional cover design. Becky Schott for the phenomenal cover photo of the *Boland*. Hopefully, people like the book as much as they do the cover! Thank you.

Joe and Heidi Porter at *Wreck Diving Magazine* for publishing my articles. Joe seems to like all the historical and archival stuff I find which has never been published. I have plenty more, Joe. Get a subscription.

Alex Gibson, my publisher and editor at Dived Up! Thanks for taking a chance on a Yank across the pond. The quality, guidance, and layout are unparalleled. Let's do another one.

Introduction

I conducted my first open water dive in Lake Erie on my 17th birthday, in September 1996, to the wreck of the tug *Admiral*. It was a surreal moment on that anchor line descending to the depths. Moving down, hand by hand, was like going back in time. On this dive, I was headed back to December 1942.

While World War II was waging on two fronts, another war was being fought on the second smallest and shallowest of the Great Lakes. The *Admiral* and its consort, the tanker barge *Cleveco*, were fighting for their survival against a fierce winter storm. It was a battle they fought valiantly, but one they would ultimately lose. Thirty-two men lost their lives that cold December night.

As I was descending down the line, I was thinking about the quick snippet of info I had read on the tug. It was just the basics — a small paragraph or two about the storm and what had transpired. I thought there had to be more to the story. All those men lost their lives in a winter storm and there were just a few short paragraphs in a book?

As I reached the wreck I could see the remains of a tug boat that looked like it had been gently placed on the bottom of the lake. Besides the missing windows and the blown off smokestack, the wreck looked as if it was fine. It didn't make sense to me. As I peered into the wheelhouse and from inside the dark cabin saw the green Lake Erie water through the round portholes, I was hooked. It was officially over. I didn't have a chance. I was going to be a wreck diver.

Several important things occurred as a result of that dive. I will take each of them one at a time, as they deserve more than a line or two. The first was that I became instantly infatuated with wrecks and wreck diving. Diving is one thing. Diving wrecks is another. I sought out other wrecks and greater depths. I sought out specialized instruction, specific dive partners, and the best gear I could afford to accomplish my goals of diving wrecks. When one tank was not sufficient, I moved on to doubles. When two tanks were not enough I sought out mixed gas and stage bottles. I am now a closed circuit rebreather (CCR) diver.

For some divers, reaching a wreck is sufficient for their goals. Not for me — I wanted to accomplish something when I got there. My dive started when I reached it. I began planning and conducting penetration dives — researching deck plans and shipbuilding blueprints. If I wanted to make my way to the engine room of a wreck, and not just swim through an open pilothouse door, I wanted to know how to get there. I would research where the hallways were, how to traverse corridors, etc.

I have always been a proponent of progressive penetration. As this is not a dive manual I will not labor the point, but suffice it to say progressive penetration is one way to learn how to dive inside a wreck. In lieu of going to the engine room on the very first dive to a shipwreck, I would make multiple penetrations over several dives so that I would be comfortable with the surroundings before advancing further into the belly of the wreck. Thus, in theory, if something did happen inside (out of air, lost/severed line, broken light) I would know the interior so well that I could find my way out and not be beholden to that one thing I was lacking, or missing, or that had failed.

When I wanted to stay longer I started decompression diving. I guess you could say it was the desire to explore that made me push my own limits to get to the depths and wrecks I wanted to see. For me, wreck diving was not just swimming around the wreck, but was a combination of diving deep, conducting decompression and making penetration dives. Like Poseidon's trident — wreck diving was, and still is, three-pronged for me.

The second thing that was borne out of that very first wreck dive, which I would argue was the most profound result, is the curiosity factor. It all started with two words — how and why. How did this wreck get here? Why did this wreck end up here? It progressed from there. What happened at the surface that was so terrible that this ship could not survive the Inland Seas? Why were there no survivors? Thus began my quest to research the answers. They are simple questions with complicated and fatal conclusions.

Research is difficult. It is not simply typing something into your favorite search engine or reading an online encyclopedia or wiki. It is done in basements of libraries. It is done on microfilm reels. Yes, I wrote microfilm reels, and yes, I have some in my office. It is done by pulling boxes and wading through thousands of pages at the National Archives looking for that one page or photo. It is corresponding with archivists, research librarians, and historians. It is pulling newspapers from the late 1800s at the Library of Congress. It is learning the process for and then submitting Freedom of Information Act (FOIA) requests. It is dealing with the intricacies of the government. It is learning how and where information is stored. It is writing to the US Coast Guard Historian's Office. Most importantly, it is time. Good research takes time. The fictional character Indiana Jones said "You want to be a good archaeologist … you've got to get out of the library!" Well, if you want to be a shipwreck researcher, you have to get into the library.

The wreck dive for me is not about the wreck itself — the pile of wood or steel in the depths of a body of water. It is not about the completely intact vessel sitting as if a child placed it on the bottom of a bathtub with no evidence of damage. It is not about the casual swim past the twisted steel and cables. It is much more than that for me. It is about the story of what happened to all the crew and passengers. It is the story of what happened the night the ship sank. It is the lost history. It is what the captain must have felt knowing he would go down with the ship. It is what the passengers must have experienced when they saw their ship catch fire, knowing no help would arrive. It is knowing what that parent might have felt when they first realized the ship was going down and had a choice of whether to drown or burn to death while they were holding their infant child in their arms. It is this human factor that makes the wreck dive complete for me. I cannot think of one wreck dive I completed without at least some sort of research conducted prior to jumping in the water. Most of the time I would spend countless hours researching a wreck's complete history before attempting a dive to her. My experience was made whole by doing so.

In my first book, *Shipwrecks of Lake Erie Volume One*,[1] these stories are plentiful. Each chapter is chock full of remarkable feats of bravery and harrowing tales of survival. I believe it was these stories — many of which had never been published before — that ultimately made the book very popular. The first two printings sold out in nine months.

The feedback I received from that book was overwhelming. I received nearly identical reactions at the presentations I gave throughout the country at various dive shows and dive clubs. Hearing the tale of John Kane, a first mate aboard the *Cortland* sleeping off his 21st birthday celebration, suddenly awoken by a giant sliver of wood piercing the left side of his face and exiting his right cheek while splitting his tongue into three and breaking his jaw, stuck into people's minds. They would never look at a shipwreck in the same way again.

I am most often asked when I will write *Volume Two*. Considering *Volume One* took six years of research and an entire year to write, suffice it to say I am working on it. However, several of the wrecks I was planning to include in the second volume are presented here. This is not officially *Volume Two*,

---

1 Forthcoming 2nd Edition, Dived Up Publications (2019).

but a separate book written in the same style which encompasses the deepest wrecks in Lake Erie. These 19 wrecks are well beyond the limits of sport diving and require very specialized training, gear and experience to reach them.

The stories of death and survival, which have had such a profound impact on me and my readers, are once again front and center in this volume. These stories need to be told in order to pass them on to future generations. If they became forgotten to history — that would be the ultimate tragedy.

I hope you enjoy the stories as much as I enjoyed researching and writing them.

Dive … into Great Lakes history!

Erik Petkovic
February 2018 — somewhere over Alaska

1 Acme

Diver swimming along one of the hogging arches

History

The tale of the *Acme* is set in a unique time period in United States history. When she was built in 1856, the country was trying to establish its identity. It was rapidly expanding. However, turmoil was in the not-too-distant future, which would ultimately tear the country apart. At the time she foundered in 1867, the country was once again trying to establish an identity, but this time for different reasons. The nation was recovering from a president's assassination and the end of the Civil War.

The *Acme* presented in this chapter is not to be confused with the tug *Acme* which foundered in 1902, or the *Acme* propeller which sank in 1893. These are two different ships entirely. As I suggested in the *Introduction*, shipwreck research can be challenging and oftentimes confusing — it is made even more so when there were multiple ships with the same name. It is further complicated when ships with the same name sailed at the same time on the same lake. Thus, it is imperative to make sure the research is accurate.

The *Acme* presented in this book was built in 1856 in Buffalo, New York by George Hardison. This early propeller of 762 tons and measuring over 190 feet with a beam of 33 feet was constructed for carrying both passengers

Ship

Official Number 297
Type Hogging arched propeller
Built 1856
Dimensions 190′10″ x 33′3″ x 12′9″
(58 x 10 x 4 m)
Tonnage 762 gross tons
Power Steam Engine
Builder George Hardison, Buffalo, NY
Owner Western Transportation Company, Buffalo, NY
Previous Names N/A
Date of Loss 5 November 1867
Cause Storm
Lives Lost 0
Location GPS 42 36.607 -79 29.853

Dive details

- **Max Depth** 130 feet (40 m)
- **Visibility** 20–60 feet (6–18 m)
- **Water Temp** low 40s°F (4–7°C)

Safety

There is a possibility of currents on this wreck. Be aware of braces and structural supports sticking out from the hogging arches when swimming nearby. Even in summer the water temperature is in the low 40s, so a drysuit is essential.

★ Deep

and freight for the People's Line between the upper lakes and Buffalo. The 13 June 1856 edition of the *Buffalo Daily Courier* stated the *Acme* was a "handsome craft".

The *Acme* had a tumultuous beginning. Just one year — almost to the day — after being launched, she went aground at Point Pelee, Ontario. If you look at it from the perspective of "glass half full" then the *Acme* was lucky for being able to get off the Point without much damage — only $100 worth. The Point has wrecked many hundreds of ships over the years and most were not so fortunate.

Two months later the *Acme* would endure her first tragedy. At approximately noon on 12 September 1857, while at dock in Grand Traverse Bay, one of her deckhands — Jack Falkonet — stepped off the deck and drowned.

Two months later the *Acme* would get her first taste of a Lake Erie November gale. Fortunately, she only lost her smokestack and anchors. The damage only cost $1,000 and, most importantly, no lives were lost. Unfortunately, this would not be her only run-in with the infamous gales of November.

After an uneventful 1858, the *Acme* resurfaced in the news in November 1859. While she was passing the south pier in Milwaukee, Wisconsin, she struck the bottom, smashed her rudder and suffered other hull damage. She was heavily loaded. She was ultimately towed back to port and repaired.

The 1 November 1861 edition of the *Buffalo Daily Courier* contained a small article entitled "A Sad Catastrophe" in which it described the *Acme*'s involvement in a rescue on Lake Michigan. On 25 October, she was 25 miles out of Milwaukee when they came across the wreck of a small fishing vessel named *Grape Shot*. The men aboard the *Acme* noticed two men "clinging for dear life" to the overturned vessel. Both men were pulled from the water "and though nearly dead from fatigue and exposure were finally resuscitated". One of those rescued was the captain — Peter Colberg. He explained that there were originally five men in the water, but prior to the *Acme*'s arrival, "three of the five had dropped off through exhaustion and were drowned". A sad tale indeed.

Over the winter months of 1863–4, the ship was almost entirely rebuilt. Later that year, again in November and during the middle of the Civil War, the *Acme* would once again test her fate with the Great Lakes. The first time she lost her stack and anchors, this time she would develop a leak. Although the only damage done was to her flour and grain cargo, her next run-in with a November storm would be her last.

In early June 1866, at approximately 0100 hours, the *Acme* had just finished crossing Lake Erie loaded with merchandise destined for Chicago. She had just entered the Detroit River when her lookouts spotted the 278 gross ton steamer *George W Bissell*. The *Bissell* was bound down with a load of lumber for Cleveland, Ohio.

According to the *Acme*, the *Bissell* was about a point-and-a-half on her starboard bow — showing her green (starboard) light and her bright light. Those on the *Acme* claimed the *Bissell* then struck her "head on, starboard midships".

Readers might not be surprised, but the crew of the *Bissell* stated something entirely different. They claimed they saw the red and green lights of the *Acme*, but as the *Bissell* approached the *Acme* inexplicably turned off both her lights. This does not make sense. Why would a ship turn off her lights to avoid a collision?

They agreed a collision took place but, while the *Acme*'s compliment maintained that the *Bissell* collided with her, the opposite crew stated that the *Acme* struck their port bow. The *Acme* suffered minor damage to the tune of a whopping $100. The *Bissell*, on the other hand, sank in the river. She was later raised.

Sinking

On Tuesday 29 October 1867, the *Acme* departed Chicago. Her destination was her home port of Buffalo. She was loaded with 2,274 barrels of flour, 236 barrels of beef, 334 green hides, 26 bundles of sheep pelts, 70 bags of timothy seed, 200 tiercin lard, 958 tiercin beef, and ten bundles of oil.[1] There were no passengers aboard — only 28 crewmen.

On Sunday 3 November, the *Acme* passed through the Detroit River and entered Lake Erie at approximately 1030 hours. The following morning at 0730 hours, she passed Point Pelee at which time the wind was out of the south and the water was "smooth". She was accompanied by the steamer *New York* — also destined for Buffalo.

At approximately 1600 hours on 4 November, the *Acme* changed course, at which time "smoother water was reached, and a lee under the south shore". When the *Acme* was approximately six miles off the shore of Ashtabula, Ohio, the wind became fierce. Captain William Dickson stated she "was tight up to 1130 Sunday night" when the wind suddenly shifted to west-northwest, which made for a "very heavy cross sea". The *Detroit Post* later wrote:

> "The boat commenced to labor badly in the trough of the sea, shipping large quantities of water. At midnight she was rolling terribly, and the exhaust pipe gave way wholly, and the main steam pipe partially, being probably twisted off by the rolling vessel".

The *Acme* was in trouble. Throughout the early morning hours of 5 November, she valiantly battled the wind and waves. She had encountered a famous November gale and was fighting for her life. Water was constantly "running all through her" until the point at which "it was impossible to get more than 20 pounds of steam". The men were working hard. Each was either manning the pumps or throwing cargo overboard — anything to try to keep the ship on the surface.

Captain Dickson made one final attempt to gain control of his ship. He needed to bring her bow into the wind, otherwise the *Acme* would continue to be pounded. He thought of a daring plan to simultaneously

---

1   Tiercin was an Old English unit of measurement still used around the Great Lakes at this time. It was equivalent to partial barrels.

drop both anchors while trying to raise the jib, thereby turning the vessel into the wind. It did not work. The *Detroit Post* reported what happened next: "The water soon reached the fires, the wheel stopped, and the boat again swung into the trough". A ship without power in a storm is doomed. Once a steamship's fires are put out it is only a matter of time.

Somehow throughout the night and early morning hours the crew was able to keep the prone vessel afloat. They were so close to shore, yet so far away. However, at approximately 0800 hours, despite the valiant effort, it became apparent to Captain Dickson that the *Acme* could not last much longer. He ordered his men to lower the boats.

At approximately 0858 hours, as the water reached the main deck, the men abandoned ship. As they rowed away in their lifeboats, the *Acme* sank stern first into deep water. The 28 men rowed the 15 miles to shore — all lived to tell the tale.

The storm which the *Acme* sailed into proved disastrous, not just for her, but for multiple other ships as well. The schooner *W W Arnold* was thrown up onto the breakwater at Buffalo. The bark *P S Marsh* lost two sailors when they were blown overboard by the driving wind. The brig *General Worth* foundered off Barcelona, New York. The schooner *Supply* was blown ashore at Port Colborne, Ontario. An unknown vessel's mast was seen sticking out above the surface near Sturgeon Point, with no sign of the crew. They were all feared dead. Another unknown schooner went ashore near Black Rock. The schooner *Mountaineer* went ashore and was badly damaged. In addition, the schooners *Atlanta* and *George Worthington*, and barks *Queen City* and *Corning* all suffered major damage en route to Buffalo and were each lucky to escape the storm with their lives.

The 5 November morning edition of the *Buffalo Daily Courier* wrote a stunning, yet concise, description of the storm:

> "A few hours will, in all probability, bring us fresh tidings of disasters, for seamen report that they had the roughest time on Lake Erie during the gale that they ever experienced".

As was typical for this time period, the aftermath of the accident was assigned to the local coroner. Today, coroners are responsible for the recovery of the dead and the subsequent analysis as to why that individual passed — determining the cause of death. One hundred and sixty years ago, the local coroners not only retrieved the bodies, they also supervised the recovery of any cargo in shipping disasters. The loss of the *Acme* was no exception.

Coroner Richards employed "a gang of men" to recover the deck load of the *Acme* which consisted mostly of barrels of lard, flour, and oil. Obviously, these barrels were stored on the deck of the ship. As long as they did not develop a leak from any damage they might have incurred while being tossed around prior to the ship foundering, they would have floated. This flotsam — which was described as being "afloat along the lake coast for several miles" — is what the coroner was in charge of securing. In all, Coroner Richards and his men worked tediously in heavy seas and recovered 600 barrels of flour, 170 barrels of lard, and a single barrel of oil. They had taken an additional 350 barrels of flour, but those were carried away when the wind suddenly shifted on the coroner's men.

Initial news reports stated the *Acme* sank in approximately 200 feet of water. However, over the coming days, those reports were updated to a more accurate depth of 130 feet.

One newspaper article from the *Buffalo Commercial Advertiser* dated 9 November 1867 contained something very interesting. In addition to the depth and giving a description of the cargo in the hold, the paper printed that "There is also a considerable sum of money in the safe" and "efforts will probably be made to raise the vessel".

I cannot find records of any attempt to raise the *Acme*. There was one newspaper article from 1871, which stated that the wrecking steamer *Rescue* was going to attempt to recover a number of wrecks in Lake Erie, including the *Acme*. However, there is nothing further. During this time period, multiple ships were being salvaged and raised on the lake. In fact, in 1853, the steamship *Erie* was raised and towed to Buffalo (see *Shipwrecks of Lake Erie Volume One*). Two years later, multiple attempts were made to raise the *Atlantic* (see chapter *3 Atlantic* in this book for a full account). However, they all failed. I was also not able to locate any record of divers salvaging anything on the wreck — including the safe. So, if there was a safe which might have contained a "considerable sum of money," it most likely is still there.

Diving the Acme

The wreck of the *Acme* rests at a maximum depth of 130 feet. I have included this wreck in the book even though most of it can be viewed at an approximate depth of 110 feet. She is almost completely buried in the mud, however, she is worth diving. The highlight is viewing the still intact hogging arches which cover approximately half of the wreck. There is a mooring line which is attached to a concrete block upwards of 50 feet away from the port hogging arch. However, there is a jump line connecting the two. A swim between the two arches is highly recommended. This is a unique chance to view early steamship construction up close.

    The superstructure was blown off during the storm, so the deck is open from stem to stern. However, the giant engine protrudes up through the deck and is worth inspecting. The stern is mostly buried and not much can be seen with the exception of the rudder post. The bow is in much better shape with a large windlass present and the stem can be seen too.

> A unique opportunity to see early steamship construction. I suggest a swim along the intact hogging arches.

2 Andrew B

The dredge barge at work

History

The *Andrew B* was constructed by Port Weller Dry Dock whose beginnings in 1946 were in repairing vessels. However, after World War II, the company began lengthening and constructing too. The 388 gross ton dredger was the eleventh vessel built by the company in 1958. She was later purchased by the Canadian Dredge and Dock Company.

After serving the Great Lakes for nearly 40 years, doing the mundane work dredgers do, she would be lost in one of the deepest parts of Lake Erie. The *Andrew B* had completed a dredging job at Bruce Mines, Ontario in Lake Huron. After a brief stop in Port Stanley, Ontario, she continued on her journey towards her ultimate destination outside Toronto. She was in tow of the tug *Offshore Supplier* when they encountered a fierce storm. The towline broke and the dredger was left to fend for herself without any power or way to navigate. After being tossed around by the heavy seas, the *Andrew B* flipped over and went straight to the bottom on 8 November 1995.

Ship

Official Number 189887
Type Dredge barge
Built 1958
Dimensions 120' x 50' x 8'
  (37 x 15 x 2.4 m)
Tonnage 388 gross tons
Power Towed
Builder Port Weller Dry Dock, Port Weller, ON, Canada
Owner Canadian Dredge & Dock Company, Toronto, ON, Canada
Previous Names N/A
Date of Loss 8 November 1995
Cause Storm
Lives Lost 0
Location GPS 42 28.796 -80 04.249

Dive details

- **Max Depth** 185 feet (56 m)
- **Visibility** 50 feet+ (15 m+)
- **Water Temp** upper 30s to low 40s°F (3–7°C)

Safety

Extremely dark. Can be disorientating due to the way the wreck lies. Entanglement hazard — hanging cables.

★★★ **Trimix**

Diving the Andrew B

Even though she is one of the most recent wrecks in Lake Erie, the *Andrew B* is not dived that often. Resting at a depth of 185 feet in dark water, she is visited by few. What complicates the matter even more is that the wreck is hard to navigate and can be confusing.

The derrick landed on her side. As a result, her deck is perpendicular to the bottom, with the top rising 40 feet to a depth of 145 feet. Remarkably, the crane and A-frame are still firmly attached and can be examined.

Most of the derrick is covered in zebra mussels (*Dreissena polymorpha*). This invasive species was first recorded in the Great Lakes in 1988. A year later they had colonized anything in Lake Erie that they could attach themselves to. Zebra mussels thrive there due to it being the shallowest and warmest of the Great Lakes. Although invasive, they have increased visibility by filtering the water. They now completely cover most of the wrecks in the lake. It is difficult to spot any exposed wood.

> Zebra mussels completely encrust most of the wreck. However, red paint is visible on some parts.

3 Atlantic

An early sketch, from 1849

Background

The entire story of the *Atlantic* is not one that can be told within the confines of this book. It is not a matter of not wanting to tell it, rather there is not enough space to do it justice. An entire volume, or possibly two, would be needed to write the complete history of this palatial steamer. One whole book could be dedicated just to the controversy and seemingly never-ending legal battles that ensued after the wreck was re-discovered and salvaged in the 1980s and 1990s.

Divers, historians, salvagers, archaeologists, those seen as opportunists or con-artists and even the general public all had passionate feelings and words for anyone who was not on their side. Then there was the invisible line that bisects Lake Erie. You know, that line that separates Canada from the United States. Even though the line cannot be seen, it is very real. An American ship in Canadian waters. Questions of ownership, speculation about untold treasures, accusations of bias, challenges to legal authority, and court systems in two different countries all complicated the matter for the admiralty and maritime lawyers.

There has been endless drama caused by artifact recovery headlines,

Ship

Official Number None assigned
Type Side-wheel steamer
Built 1849
Dimensions 265'7" x 33' x 14'6" (81 x 10 x 4.4 m)
Tonnage 1,156 gross tons
Power Low pressure walking beam engines
Builder John Wolverton, Newport, MI
Owner Samuel and Eber Ward, Newport, MI
Previous Names N/A
Date of Loss 20 August 1852
Cause Storm
Lives Lost 250–300
Location GPS 42 30.620 -80 05.086

Dive details

- **Max Depth** 165 feet (50 m)
- **Visibility** 20–70 feet (6–20 m)
- **Water Temp** low to mid 40s°F (4–7°C)

Safety

Heavily silted. Very dark. Wreck can be confusing as it is largely collapsed. Mooring line is usually tied to the walking beam engine amidships between the two paddlewheels.

★★ Tech

lawsuits, the debate between marine archaeologists who believe artifacts should remain on the wreck, salvagers who recover for profit, and others who believe that artifacts should be recovered but preserved in museums for everyone to see. However, the most important part of the *Atlantic* story is that the loss of the steamer ranks as the fifth worst maritime disaster in the history of the Great Lakes, in terms of loss of life.

In lieu of telling the entire story or presenting a very watered down, light-on-detail version crammed into a few paragraphs, I will give a lengthy — albeit shortened — account of the *Atlantic*. I will focus on a long forgotten and seldom written history of the early salvage divers, the recovery of the famous American Express safe which was on board when she foundered, a detailed description of some rediscovered tales of death and survival, and a very public showdown about the quality of life preservers aboard the vessel.

Construction

Built in 1849 in what is now known as Marine City, Michigan by shipbuilder John Wolverton for Samuel and Eber Ward, the *Atlantic* was a spectacular ship. Operated by the Michigan Central Railroad Company, she measured 265 feet long with a beam of 33 feet. She had a gross tonnage of 1,156.

The wooden steamer was driven by two firebox boilers built by Hogg & Delameter. Each measured 10 feet by 34 feet and powered her side-wheels. Her massive low pressure walking beam engine had impressive five foot diameter cylinders with an eleven foot stroke. All this power gave her an average speed of 15 miles per hour. However, she was capable of higher velocity. On 7 June 1849, the *Atlantic* set a record for the fastest crossing of Lake Erie from Buffalo to Detroit, taking 16 and a half hours.

Even while being constructed the *Atlantic* was very popular with the press of the time. The *Buffalo Daily Courier* stated:

> "The *Atlantic* is an ornament to our lake marine, which already embraces many of the most splendid vessels on the American waters. In speed and accommodations for the traveling public, we are confident she will prove in no respect inferior to them".

While her painting, trim, upholstery, and other decorations came out of Buffalo and other parts of New York, her crockery was specially made in Liverpool, UK, and her official plate came from Birmingham, UK. The ship was even finished with beautiful, handcrafted rosewood furniture of "exquisite workmanship". The 26 May 1849 edition of the *Buffalo Commercial Advertiser* stated: "The *Atlantic* is built with the modern improvements and is another beautiful specimen of marine architecture". One publication even characterized the furniture and furnishings as "regal magnificence".

Once completed, the ship, which most thought far exceeded everything on the Great Lakes in terms of grace and speed, would run her regular route between the ports of Buffalo and Detroit — departing each city on alternate mornings. Her 85 staterooms would be routinely filled, as would the 150 cabins. Even though the grand steamer was always booked to capacity, yet more steerage and deck passengers were crammed into every available space — as long as they paid one dollar.

Loading

There were no loading regulations for vessels during this time on the Great Lakes. There were no signs stating the maximum number of passengers allowed on board. There were no customs or port agents to verify cargo or count passengers. The sole authority as to what was loaded, how much, and when it was finished was the captain. Of course, as one could guess, the driving force behind the loading of the ship was money.

The captains of this time were under an incredible amount of pressure to maximize profits. This often meant taking on excess cargo to the point that ships were overloaded. Many a ship met her demise on the Great Lakes due to being grossly over burdened with cargo. Another way to maximize financial gain was to take on extra passengers. Any space that was available on the ship — whether above or below deck — was a potential berth for a paying customer. At this time in American history, these boat spots often went to immigrants who were heading west in search of a new life. Those who filled up the *Atlantic's* vacant spaces were no exception. They were Europeans in search of a better life in America. Of course, none of these individuals were ever recorded on the official passenger manifest.

Advertisement for the *Atlantic*, 1852

On 19 August 1852, the *Atlantic* was alive with noises and the hustle and bustle of loading a ship. Imagine the scene in the movie *Titanic* — porters carrying steamer trunks, people dressed in their finest, some of the women holding parasols to shade themselves from the sun, etc. — that is exactly how it was at the *Atlantic*'s berth in Buffalo, 60 years earlier. She was being prepared for her regular run to Detroit. The ship was loaded with cargo including United States mail, approximately 1,000 guns, a carriage, scores of crates and hundreds of steamer trunks containing all of their owners' worldly possessions including gold, jewelry and money.

The earliest to board the ship were the first class cabin passengers. They each paid a handsome ten dollars for their ticket. Once aboard, most went to the grand salon to relax, smoke a cigar or grab a drink. Others listened to the orchestra. Some toured the decks and watched as the steerage passengers boarded. These overwhelmingly Norwegian and German immigrants were kept separate from the first class and cabin passengers at the dock.

According to the Michigan Central Railroad Company, they sold 326 tickets for the trip to Detroit — this included first class and other cabin passengers throughout the various decks. Captain Pettey later stated there were 600 people aboard. If one deducts the total number of tickets sold and the 125 crew members, that leaves approximately 150 steerage passengers. The exact number will never be known as the manifest went down with the ship. This coupled with the fact that ships' manifests never truly showed the correct number of paying customers anyway in order for there to be no record of overloading a ship. Most historians agree there would have been between 510 and 600 passengers and crew aboard when they departed.

Fateful journey

At approximately 1730 hours, the *Atlantic* pulled away from Buffalo. If all went well, Captain J Byron Pettey would have the ship in Detroit at approximately 1000 hours the following morning. The weather was warm. The seas were calm. There was slight fog along the top of the water, but nothing that would impede visibility.

At midnight, the *Atlantic*'s second mate, James Carney, reported to the wheelhouse and took command of the watch. Captain Pettey retired to his cabin for the night. At the helm was Morris Barry.

Meanwhile, the *Ogdensburg*, operated by the Northern Transportation Company, had departed Cleveland, Ohio, traveling northbound. The *Ogdensburg* was a 352 ton, two-deck steamer measuring 275 feet in length. She was bound for Oswego, New York via the Welland Canal.

At 0200 hours, Carney and Barry saw the Long Point Light approximately two miles distant. They were traveling west by southwest.

On board the *Ogdensburg*, First Mate DeGrass McNell was in the wheelhouse on lookout. McNell saw the lights — red light with two white lights below — of a westbound ship approximately three miles off the *Ogdensburg*'s starboard bow. Due to the distance between the ships and their courses, McNell believed the *Ogdensburg* would pass at least one-half mile south of the westbound ship.

At approximately 0315 hours, Carney saw the lights of an unknown ship. He ordered a slight correction to port. He was comfortable doing so as the lights he saw on the northeast bound ship were dim, which led him to believe the ship to be several miles distant. While the crew of the *Atlantic* was comfortable with this course change, aboard the *Ogdensburg*, McNell was shocked and stared in disbelief as the *Atlantic* changed course to bring her across the *Ogdensburg*'s bow!

> One of the worst maritime disasters in Great Lakes history

McNell immediately ordered the *Ogdensburg*'s engines stopped, but it was too late. The fates of hundreds of people were already sealed. McNell went to ring the bell, but then remembered it was inoperable. The *Ogdensburg*'s wheelsman then turned the ship hard to starboard in an attempt to avoid collision. McNell ordered the engines reversed on the telegraph. McNell shouted out to sea hoping the *Atlantic* would hear him. For ten agonizing minutes, the *Ogdensburg* moved swiftly towards the oncoming ship even though her engines were stopped, then reversed. She was running full throttle and had too much forward momentum to be able to stop in such a short distance.

Just as Carney and the rest of the crew in the *Atlantic*'s wheelhouse saw the other ship out of their port window, the *Ogdensburg* smashed into her forward of the port side-wheel — between the forward gangway and the wheelhouse. The *Ogdensburg*'s bow sliced almost halfway through the *Atlantic*'s maximum beam. The damage sustained by the *Atlantic* extended far below the waterline.

The two ships separated very quickly as the *Atlantic* was running wide open and never slowed prior to the collision. Carney immediately ordered the *Atlantic* to head for shore, which was approximately four miles distant. Captain Pettey reported to the wheelhouse. They believed the only way to save the ship was to beach her.

Carney immediately descended below decks to triage the damage. The forward steerage section he checked was dry, but he was unable to access the lower decks as the hatches would not open. He then went to the boiler room, where mass chaos was breaking out. Carney later stated "water was rushing in, in torrents". He heard the chief engineer screaming orders for the workers to stuff bedding and anything else they could find into the gaps to slow the water. They needed to buy time, otherwise, the water would rush in and put out the fires in the boilers.

Meanwhile, in the hold, the crew was attempting to move the cargo to the starboard side. They were hoping they could slightly lift the port side damage above the waterline.

The *Ogdensburg*'s engines were stopped ten minutes prior to the impact. She was so powerful and had so much forward momentum that she finally stopped nearly three miles beyond the site of the collision! As soon as the crew ascertained that the damage would not cause the *Ogdensburg* to sink, they turned and headed for the *Atlantic*. The *Atlantic* continued racing towards shore. She only made it a half mile before the fires were extinguished by the water flooding the boiler room. She was now dead in the water. The fact that she didn't sink immediately after being nearly cut in half is a testament to her construction.

While the *Ogdensburg* was racing towards the scene of the collision, all hell was breaking loose on the *Atlantic*. Captain Pettey gathered some crewmen and started working on lowering the lifeboats. The first problem was that there were only three of them. The second complication was that one of the boats fell overboard, landed upside down in the water, and was rendered completely useless. The third issue was that while attempting to ready one of the boats, Captain Pettey fell eleven feet and landed on a lower deck. He suffered a massive gash on his head and was lucky to survive. The fourth problem was that when the last two lifeboats were launched — much to the horror of the passengers — they were filled by the crew, who quickly rowed away!

While Captain Pettey, First Mate Carney and the wheelsman were attempting to save the ship — thereby saving those on board — some of the crew were not doing their duty in assisting the terrified passengers. Whether it was due to incompetence or a lack of training or if they were frozen with sheer terror, there were multiple reports of crew members not doing anything to help the passengers. It did not help that hundreds of passengers watched as the crew took the lifeboats for themselves.

As the *Ogdensburg* was still heading towards the stricken ship — it would take nearly an hour for them to arrive on scene — complete pandemonium had overcome the *Atlantic*. It is presumed that the 150 men, women, and children who occupied the forward cabins below deck all drowned. Those who were sleeping in the cabins at the point of impact were crushed to death if they were lucky. Otherwise, they would have succumbed to a nightmarish drowning death while trapped by debris, unable to escape.

Engraving of the collision between the *Atlantic* and the *Ogdensburg* published in *Gleason's Pictorial* on 11 September 1852

Many others on board had no chance to escape as some of the interior bulkheads had shifted so severely due to the impact that doors could not be opened. Jacob Nash was in his stateroom along with his wife and their family friend at the time of the collision. He attempted to open the door but it would not move. Water was entering the cabin. Jacob broke the window above the door and pulled himself out of the cabin, which was rapidly filling with water. He was bleeding from the glass which had sliced his arms and body. He grabbed a nearby chair, placed it against the door, stood on it, and shouted through the broken window for his wife to climb through. He was able to pull her out. It was at that point that all three realized that due to her size the friend would not fit through the window. She drowned in the room unable to escape.

One of the more harrowing tales comes from Almon Calkins — a passenger. As you read his statement try to imagine being in a berth while the one above collapses onto you. If that is not bad enough, now imagine being in your berth with the remnants of the one above in it, and then collapsing into the berth below. You never saw the man you landed on again. Then picture escaping and finding your way to the deck. You were relieved until pushed into the water by a panicking crowd, surrounded by people who could not swim and who were trying to pull you under. The screams you would have heard of people being crushed to death or of distraught people trapped — knowing they would drown — would forever be carved into your memory.

> "I perceived that the outside of my berth was carried entirely away, and why I did not fall out into the water I cannot tell. I now extricated myself and those around me from the rubbish. The water had now risen over my feet on the upper deck, and a cry arose that the vessel was sinking. I endeavored to put on my pantaloons and partially succeeded. We now started for we scarcely knew where, and I was carried by the crowd overboard. I sank with numbers clinging to me, perhaps 16 feet. I could not swim. At this instance I caught hold of a rope by means of which I regained the wreck. I next made my way under the water to the mast of the steamer. I was next pushed from the mast; there were numbers clinging to it. I now caught hold of another rope, but soon pulled, by at least four persons clinging to my legs from my hold. The marks of the grasps of those poor wretched beings are still upon my ankles".

An hour after the accident, the *Ogdensburg* re-appeared. Even though they did not hear any distress signals or bell ringing from the *Atlantic* after the collision, they decided to go to her aid anyway. If they had not tried to assist, an additional 300 people would have died. When they neared the *Atlantic*, they heard hundreds of people on board and in the water crying out for help. The *Buffalo Daily Republic* reported "the lake was covered for miles with floating fragments and persons clinging to life".

The *Ogdensburg* first moved into position next to the *Atlantic* to rescue those stranded on the stern. The *Atlantic's* lights were out and she was almost completely perpendicular to the water, with her bow totally submerged, and her stern rising high out of the water (again, think *Titanic*). After they rescued approximately 200 people from the stern, before the ship completely disappeared into the abyss below, they lowered their lifeboats and started picking up those in the water.

While the lifeboats were engaged in rescues, the *Ogdensburg* made a circle one mile in circumference around the *Atlantic,* looking for survivors. She stayed on the scene until her captain believed they had rescued "all living persons in the water". They pulled approximately 100 souls from the lake.

Aftermath

Word of what the *Detroit Free Press* referred to as an "Awful Calamity" quickly spread across the country. The 21 August 1852 edition of the *Buffalo Daily Republic* printed a story entitled "The Appalling Calamity On The Lake" in which they reported the little details they knew:

> "At the present writing we are without authentic intelligence on the extent of this last sacrifice of human life, but the accounts which we have received render it too probable that it equals, if it does not exceed, in fatality, any disaster of the kind that has ever occurred on our western waters".

Of the approximate 600 people aboard, it is believed over 300 perished. In the aftermath of one of the worst maritime disasters in the history of the United States, significant complaints arose about the lack of sufficient life preservers aboard the *Atlantic*. Many of the survivors strongly voiced their opinions that those provided were completely inadequate and would not hold anyone's weight in the water. Their frustrations boiled over to the press. The *Buffalo Daily Republic* printed the following:

> "There have been many devices of life preservers; but none that have been adopted seems to be sufficiently effective to prevent hundreds from drowning, out of a single vessel load of passengers".

The article continued:

> "Again, let every berth be provided with reliable life preservers. Not such as have to be inflated with air before they can be used, for nine-tenths of the people would not know how to inflate them, and three-fourths of the other one-tenth would be too frightened when there should be real necessity for it, to inflate them, if they knew how ever so well. Those articles of furniture which are so constructed as to have life-preserving qualities, are not where passengers know where to lay their hands on them immediately, in the night; and those stools, with tin boxes within the legs, are of very little use when they are to be found. A person in the water must have something that he can get upon and lie upon when he is on, or it will not save him. Those little

bobbing things, which turn over and over as fast as they are touched with the hand of the drowning man, would not save one out of twenty".

The survivors were talking about what was referred to as the "Ward Life Preserver" — a wooden stool with a sealed tin pan containing air inside. Many stated it would not support their weight in the water. Others that it was worthless as a stool and even more worthless as a life saving device. On August 25 the *Buffalo Daily Republic* referred to seats "which unto many who trusted [them] for preservation proved Life Destroyers". There were many other harsh words for this stool which was blamed for a tremendous loss of life. Many believed if there had been another type of life preserver aboard, many of those who drowned would not have.

This ultimately caught the attention of Eber Ward — owner of the *Atlantic* — who issued the following statement in the 3 September 1852 edition of the *Detroit Free Press*:

"If the 'tin-bottomed stools' or what are called 'Ward's Life Preservers' are really valuable, it is important that the public should know it. They were very thoroughly experimented on, and their efficiency proved by about twenty boys and men yesterday in the presence of a large number of spectators, in front of the railroad depot. They jumped into the water with them and floated off without the slightest effort and remained in the river as long as they chose. But Mr Ellette H Titus, who says he is a good swimmer, found them at the sinking of the *Atlantic*, 'worse than useless' and 'mere traps'. Now I propose to bet Mr Titus or any other man $100 or $500 dollars that one man shall have his feet tied together, and also his hands tied, and jump into the river with one of these stools and remain there half an hour without any other assistance. I also will bet $100 or $500 that two men shall each have his feet tied together, and the two shall jump into the river, with one stool between them and remain in the water half an hour, or as much longer as is necessary to convince any reasonable man of the efficiency of this life preserver".

It is not known whether or not anyone ever took up Eber Ward on his wager to prove the worthiness of the stool. He vehemently defended it against any and all people who said otherwise. However, there were several reports of a single stool not floating when there were multiple adults clinging to it. One such story involved a little girl who watched in horror as her mother and aunt frantically fought over the same stool. She watched her father jump into the water to save his wife and sister-in-law. When he reached them, the women were so violent and panicked that they pulled him below the surface. The little girl looked on as all three drowned in front of her.

TERRIBLE CALAMITY!

TWO HUNDRED & FIFTY LIVES LOST!

On Tuesday morning last, about two o'clock, the steamers Atlantic and Ogdensburg came into collision on Lake Erie, in a dense fog, and the Atlantic sunk in a few minutes. She had on board an immense number of passengers, and as they were mostly in their beds, many of them found an almost instantaneous watery grave. There was a large number of Norwegian emigrants on board, who nearly all perished. The scene was one of terrible and heart-rending confusion; and the boat sunk amid the wildest shrieks from a thousand voices. All who had life-preservers or who provided themselves with chairs, settees and beds, sustained themselves in the water until picked up by the other steamer. About 250 were saved; but it is thought 250 or 300 perished! A large amount of money and property also was lost; and those saved are completely destitute and homeless,' at the same time mourning the loss of friends and relatives.

Article from the August 22, 1852 edition of the *Gettysburg Adams Sentinel*

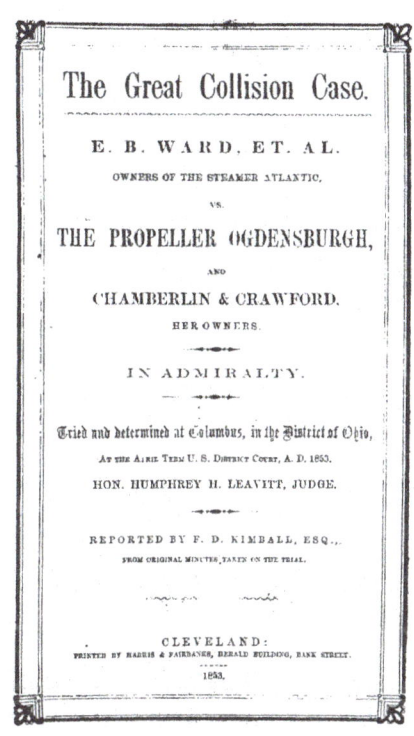

Report of the court case between the owners of the *Atlantic* and *Ogdensburg*, 1853

Legal action

As is typical in the overwhelming majority of collisions between ships, lawsuits were filed. The collision between the *Atlantic* and the *Ogdensburg* was no exception. No criminal charges were ever brought. The first civil suit was filed by the Wards in October 1852. The *Ogdensburg* immediately countered with their own suit alleging that the *Atlantic* was to blame for the collision. The case ultimately made its way all the way to the United States Supreme Court.

The two parties did not agree on anything, as is typical in legal actions. However, after much work, both parties agreed that their claims should be considered by the court which would weigh the evidence. The judge ruled the disaster was due to mutual fault and that costs — the *Ogdensburg* claimed $3,000 in damages, while the *Atlantic* claimed $75,000 in losses — should be divided between both parties. The lawyers representing the *Ogdensburg* immediately appealed. In 1859, after seven years of litigation, they ultimately lost and were ordered to pay the Wards $41,200.

Salvage attempts

Two weeks after the *Atlantic* sank, the owners Samuel and Eber Ward decided to put the wreck up for sale, as she was not insured. The starting bid was $10,000. They believed the machinery alone was worth much more. If you begin to add in all the furnishings and the ship itself, $10,000 seemed like a bargain to the Wards. There were no bids. The issue was not that no one wanted the wreck, but that no one believed it was possible to salvage something resting in 160 feet of water, let alone something as big as this steamer. It simply had never been done before.

As a result of the lack of interest, the Wards decided to sweeten the deal. The new proposal was $15,000 to anyone who could raise the ship. They received an offer almost immediately from Henry Sears, who envisaged raising her with the aid of his new *Nautilus* diving bell. The Wards asked Sears to join them on a survey of the wreck in August 1852. Unfortunately, the diving bell could not be used as it was still considered experimental at the time. It had just been constructed and had yet to be fully tested.

There was one man who was interested, not in raising the wreck of the *Atlantic* but in one particular item. That man was Henry Wells — the president and founder of the American Express Company and of Wells Fargo. He wanted to raise the ship's safe, which was located in a stateroom behind the wheelhouse. He offered $5,000 to anyone who could recover it.

Wells invited marine engineer Benjamin Maillefert, the "new submarine diving phenomenon" John B Green, Green's two diving tenders, and two additional unnamed divers from New York to join an expedition to the wreck. Green later remarked that the two New York divers had much better gear, stating that he found their armor "far superior" to that which he used.

On 24 August 1852, the group departed Buffalo for the wreck site. The New York divers, who throughout the entire trip boasted about their diving exploits and how Green could not dive deep, ironically, refused to dive once they found out the wreck rested at a depth of 162 feet. Green decided to make the dive. Wells was cautious as he thought the best divers were the ones who refused. Green, ever confident, persuaded Wells to give him a chance. However, Green only made it to 105 feet before having to stop as the air pump and his air hose were not functioning properly. The dive operations ceased due to bad weather and the need for a new hose.

The New York divers — who had bragged about diving to depths of 200 feet and staying for long periods of time and who mocked Green for "playing in shallow water and like a boy who could not swim, [and] dared not venture where it was too deep" — sat silent for the return trip back to Buffalo.

Three weeks later, on 18 September, the party returned to the *Atlantic*. Green suited up and descended to the wreck. Immediately there was a problem. Green not only located the wreck, he dropped directly into one of the still-standing smokestacks! The chances of a diver blindly jumping off a boat and dropping straight into an upright smokestack are about equivalent to seeing a shark in Lake Erie — almost nil. He was immobilized from the chest down — the only thing keeping him from being completely encapsulated inside the smokestack were his armpits. Miraculously, he was able to signal for a lift and was pulled free. He was then able to land on the ship's braces, but he had to be hauled up as the tender boat on the surface was violently rocking due to wave action. It was decided to abort.

Green's next dive was to an incredible 152 feet. One newspaper stated that this was the deepest dive ever recorded at that time. Although impossible to independently verify it is, however, a very impressive dive given the fact that it was 1852. Extraordinary as it was, Green only barely averted disaster, as his air hose burst on the bottom. He was hauled up before he lost all his gas. The group returned to Buffalo, where he constructed a much stronger and sturdier hose able to withstand the pressure of the deep water.

Green described his attainment of great depth as follows:

"I found great difficulty in moving; the water was so compressed; and with the diminutive air-pipe which we used, it was next to impossible to keep the armor inflated below the waist, and often it rose as high as the chest. The pressure was immense. The rush of blood to the head caused sparks of various hues to flash before my eyes, and I had a constant tendency to fall asleep, although the pressure on my limbs was enough to crush them under ordinary circumstances".

Less than two weeks later, on 30 September, Green headed back to the *Atlantic*. He conducted a series of eight dives. He made it to the upper deck level, but he did not know his exact location on the wreck. Although Green had made multiple dives, Wells was unsure whether a diver could successfully reach the depths required and be able to effectively orient himself enough to be able to recover the safe. He therefore decided to hire Albert Bishop to raise the wreck for an incredible $25,000 (nearly a million dollars in today's money)!

However, since no derrick of the size needed to raise the *Atlantic* existed, Bishop needed to construct it. Massive is a complete understatement. Once completed, the derrick — which was built on two steamers — measured an astounding 142 feet high and weighed a remarkable 16 tons. It had an impressive lifting power of 4,960 tons.

Multiple dives were made over several trips, but not to attempt to recover the safe. The divers were instead aiming to wrap chains around the hull of the wreck. However, they had trouble properly connecting them. The Bishop derrick, a monument to engineering in 1852, could not pull the *Atlantic* from the depths without chains secured around the hull. Before divers would be able to do this, the derrick was torn apart in a storm and sank.

Marine Cigar

In September 1853, Lodner G Phillips — a shoemaker turned engineer — had the opportunity he was looking for to test a new device that he had invented, called the *Marine Cigar*. This submarine tapered just like a cigar, hence the name. It weighed approximately eight tons and measured 40 feet in length and four feet in diameter. It was powered by a hand-crank that operated its two-bladed propeller. The United States Patent Office provided Lodner with Patent #9389 for his steering design.

There have been multiple reports over the years since that this early submarine was used in an attempt to locate and raise the *Atlantic*. There is no known documented evidence, writing or report which lends credence to this actually having happened. However, there were two newspaper reports which may have propelled the belief that Lodner was looking for the *Atlantic*.

The first article appeared in the *Detroit Advertiser* entitled "A Sub-Marine Propeller". It gave a detailed description of the submarine:

"We saw yesterday at the Railroad Freight House a curious looking structure of wood and iron, shaped something like a pear, only about 20 feet long, with a little propelling paddle-wheel at one end, and an iron flanged steering paddle at the other. On the sides were small bull's eyes windows, filled with very thick glass. The machine, we were informed, is Phillips' Sub-Marine Propeller, and came over by the railroad from Michigan City, on its way to pay the *Atlantic* a visit".

A second article appeared in the *Buffalo Daily Republic* stating:

"We noticed Phillip's Patent Diving Boat on the Michigan Central Dock this morning having been brought down by the steamer *Ocean*. The owner of the boat intends making an attempt to reach the *Atlantic* in a day or two with his new invention".

That article, from 15 October 1853, is the last one about Phillips and his submarine. What happened next is pure speculation. However, based on the memoirs of his nephew and a story passed on by his son, it is surmised that Phillips' *Marine Cigar* foundered in Lake Erie. These family members stated that, in 1853, Lodner wanted to locate and salvage a wreck in the lake, but did not specifically state that it was the *Atlantic*. In order to test the submarine, Phillips supposedly took it down to a depth of 100 feet with a load of passengers. While at depth, the submarine developed a leak. Upon surfacing, all passengers were offloaded — including Phillips. The *Cigar* was then put back down on the bottom without anyone aboard. This point is important as other literature states that Phillips went down with the submarine, but that is not true. Supposedly, everyone saw a giant bubble of air. Lodner was not able to lift the submarine back to the surface. She had a catastrophic leak, filled with water and was lost to history.

Somehow, over the years the legend grew that the submarine was lost while over the *Atlantic*. Some have even stated that she literally came to rest on top of the wreck. This, however, is not true. Even though she is not dived all that often, there have been enough dives over the years that someone would have noticed a submarine lying across the deck.

There have been numerous attempts to locate the *Marine Cigar* using side scan technology. No submarine has appeared on those scans. So, unless she is completely buried beneath the bottom, she is not there. The story of the submarine may be a case of what J R R Tolkien wrote in *The Lord of the Rings* — "And some things that should not have been forgotten were lost. History became legend. Legend became myth. And … passed out of all knowledge".

UNITED STATES PATENT OFFICE.

L. D. PHILLIPS, OF CHICAGO, ILLINOIS.

SUBMARINE EXPLORING-ARMOR.

Specification of Letters Patent No. 15,898, dated October 14, 1856.

To all whom it may concern:

Be it known that I, LODNER D. PHILLIPS, of the city of Chicago, in the county of Cook and State of Illinois, have invented certain new and useful Improvements in Submarine Armor or Exploring Apparatus; and I do hereby declare that the following is a full, clear, and exact description thereof, reference being had to the accompanying drawings, and to the letters of reference marked thereon and forming part of this specification, in which—

Figure 1 represents a front view of the apparatus; Fig. 2, a side view of the same; Fig. 3, an upright section; Fig. 4, a horizontal section, with several other sectional views and parts of the apparatus which will be more particularly named hereafter.

In general my invention relates to the construction of a metallic vessel or apparatus so apportioned, arranged and equipped with tools or instruments for sub-marine explorations, and of proper form to admit a man in its principal cavity, and in furnishing such metallic casing with air chambers and other apparatus for the convenience and safety of the person operating the same; also suitable joints of peculiar construction to allow the free use of the legs and arms of the operator; also attachments by means of which the operator can ascend, or descend, or change his position and give locomotion to the machine without assistance from others.

a, represents a cylinder of boiler iron having the dome head *c*, of cast iron, furnished with a "manhole" which is covered with the cap *d*. This cap is bolted to the head, having a gasket of leather or india rubber in the joint. The lower end of the cylinder *a*, is fitted with a cast iron concave bottom as shown at *b*. There are two circular openings in this bottom, cast with flanges to receive the legs of the machine.

e, represents the upper and *f* the lower part of the legs.

The joints at the hips and knee are constructed on the "ball and socket," principle, as shown at Nos. 1 and 2.

The legs *e*, *f*, are of cast iron, and are suspended from the bottom *b*, by the iron straps *g*, *i*, and *k* on the outside and by the straps *l* and *m* on the inside. The upper ends of the lower straps have journals, or pivots, fitted in boxes formed in the lower end of the upper straps. These joints are shown at *h* and *j*. An iron stud or pillow block is attached to the bottom *b*, between the legs, in which the pivots of the upper ends of the straps *i*, are fitted. These pivots are concentrated with and form the axis of the socket joints. To prevent the passage of water through these socket joints I envelop them with a ring of india rubber or other suitable elastic substance, as shown at *o*. The edge of this elastic ring is secured above and below the joint by the hoops *n* and *p*. These hoops have their ends flanged and are drawn tight with a screw bolt in the usual manner.

e' is an exterior casing also of boiler iron placed at the back of the cylinder *a*, at a distance from it, and concentric to it. This casing *e'*, forms a chamber, as shown on Fig. 3 and at B on Fig. 4. This chamber is for the purpose of containing air forced into it, to any desired degree of density, and forms a reservoir upon which the operator may draw as required.

j' and *k'* are two elastic tubes inclosed in a circumscribing tube, of any desired length. The tube *j'*, is connected by the small pipe *h'*, to the side of the cylinder *a*, and opens into the air chamber B. The upper end of the tube *j'*, connects with an air forcing pump. A cock *i'*, is inserted in the connecting tube *h'*. The pipe *k'*, opens into the principal cavity, and is the escape pipe for the vitiated air, a valve being fitted to the inner end of the pipe, and is opened by the operator as occasion may require. Signals are also made through this pipe by means of the voice or discharges of air.

m', represents a cock by means of which the air is admitted from the air chamber into the principal cavity A.

b', *b'*, are two glass plano-convex lenses inserted in the cylinder *a*, and secured by the rings *o'*, *o'*. A lamp, as shown at *y*, is placed inside and near the lower lens. This lamp is furnished with a concave reflector, as shown at *s'*. The upper lens acts as a window, through which the operators may observe the objects sought, aided by the light emitted through the lower lens by the lamp. A small pipe *w'*, communicates with

Lodner Phillips' US Patent for submarine armor

Back for the safe

After multiple failures most people thought it impossible to raise the wreck, let alone anything from it. Hope had been abandoned by everyone — everyone except Green. While Green had moved on to other projects he wrote:

> "It is said that there is nothing on earth to which man will so eagerly cling, as to gold. That it occupies his thoughts by day — his dreams by night; and true to this passion, that wealthy safe at the bottom of Lake Erie had been rife in my mind…"

Green decided he did not need the support or permission of anyone else. He would go for the American Express safe on his own. The challenge was all about Green and his ego. He was a celebrated diver. He was famous for his exploits and daring dives to the steamer *Erie* (see *Shipwrecks of Lake Erie Volume One*[1] for a full chapter on that wreck) in which he recovered gold, silver and Swiss francs. He considered himself to be a celebrity. He always had something to prove to himself. Now he wanted glory.

In August 1855, three years after first diving the *Atlantic*, he chartered a vessel and took 18 men to the wreck site. It took several days for him to locate her because all the buoys he had placed three years prior to mark the location had disappeared.

Green made multiple dives to the upper decks of the wreck at 145 feet in total darkness and solitude. Wreck divers know too well the experience and feeling of solid darkness underwater, with limited to no visibility, having to feel your way around a wreck. Imagine doing that at 145 feet without a light (as they had yet to be invented), nor any safety equipment, and no back up air source. Now think about doing that dive over 160 years ago. Now put yourself in Green's position, doing that dive with the following gear, as described in the 8 September 1855 edition of the *Buffalo Daily Republic*:

> "Dressed with three pair of flannel drawers, three shirts, also flannel, three pair of woolen pants, three coats and three pair of woolen stockings, surmounted by his submarine armor; on his feet he had a pair of stodgy shoes, with a lead sole of ½ or ⅝ of an inch thick, and a belt

---
1   Forthcoming 2nd Edition, Dived Up Publications (2019).

of 80 lbs of shot around his body, to sink him, and the breast piece cannot weigh less than 50 lbs".

That is a tough dive.

What made it even more difficult for Green was that the wreck had tilted 90 degrees since he last dived it, and was also covered with ten inches of mud. The years of being submerged were starting to take their toll on the *Atlantic*. However, on 23 August, after nearly a week of repetitive, daily dives he was able to consistently land on the wheelhouse. From there he moved down to the deck. He was getting closer. He moved three windows aft to the cabin where the safe was located. Nobody had been at this exact spot since the ship sank. He reached in through the window and felt around. His hand stopped on a box. He'd found it! Elated, he recounted: "Then I cried out in my helmet, my God, I've got it! I'm a rich man! And I wept down there in the waters, I was so glad". He had done what no one thought was possible and what no one wanted to attempt because the risk of death and failure was so high.

He marked the position by tying a line to the railing opposite the cabin window. He sent one buoy to the surface and hung a second ten feet below in case the first was torn away. He took a short break after his two initial dives. On his third, he took an iron prod and a saw. After he reached the deck and located the line tied to the railing, he cut away the cabin wall through the window. Once the opening was large enough, he dragged the safe onto the deck in order to prepare it to be lifted. He signaled to be pulled up. He wanted to get some rope and a hook to complete the job everyone said was impossible. He had just done three repetitive dives to 150 feet.

Once he surfaced, his tenders signaled that the "lifting tackle" was not quite ready, so he asked them to remove his helmet, which by nature was claustrophobic. Once his helmet was off he felt a sharp pain in his legs. He then collapsed. He could not move. He was paralyzed and near death. The crew quickly sailed to Port Dover, Ontario. For two weeks he lay in his hospital bed waiting to die. Nobody knew what to do, not even the doctors, who believed he would not survive. Thankfully, after those initial two weeks, movement returned to his upper body. However, he was still paralyzed from the waist down.

It took nearly eight months for Green to be able to walk again, and even then, it was only with the help of a cane. By this time he had had enough. It was late June 1856. The only thing on his mind since he located the safe in August 1855 was bringing it to the surface. He needed to finish the job. He was going to do it one way or the other, whether he had help or not.

On 1 July 1856, Green chartered a boat and brought two divers to do the work as he was incapable of walking, much less diving. Once they reached the site it took nearly a week to drag for the wreck — trailing an anchor waiting for it to catch onto something — as the buoys were gone. The first diver down signaled to be brought up after reaching only 60 feet. Green sent the second diver down. He too signaled to be hauled back to the surface after hitting the 60 feet mark. Infuriated, Green decided to make the dive himself. Despite being unable to walk, he donned the bulky, heavy gear and took the plunge.

Unlike the other two divers, he made it well beyond 60 feet. He landed on the wreck and made his way to the deck between the railing and the cabin where he had left the safe. But it had gone.

When he was brought up to the surface, he could not speak. He fell onto the deck and was once again totally paralyzed. He suffered from what is now known as decompression sickness — the bends. Green later described in great detail what he felt the exact moment the bends crippled him:

> "I had sat but a moment when a sharp pain shot like lightning through my lower extremities, and the next instant it went through my whole system, so prostrating me that I could not move a limb, or even a muscle".

While in the hospital he learned what had happened. His rival, a submarine diver — to give him the full title used at the time — named Elliott Harrington, recovered the American Express safe just four days prior to Green's dive. Green was incensed. He accused Harrington of following the buoys he had left attached to the railing. He believed Harrington stole it from him. To make matters worse, the small safe (measuring only 28 x 16 x 18 inches) contained $5,000 in gold, $31,000 in paper currency, and six watches — well over a million dollars' worth in today's money.

Even though Green would not become rich as he had hoped and dreamed, he did receive some satisfaction regarding the contents of the safe. Harrington and his three men had agreed to split the treasure four ways. After discovering that $3,000 of paper currency was ruined, each man would receive approximately $6,000. Given that this was in 1856, it was still a considerable amount of money. However, the men were visited by an American Express Company lawyer who offered them a total of $7,000 for the safe and all of its contents. If they refused to hand over the safe, the other option would be a costly lawsuit which the company was willing to file. Wisely, Harrington and his men settled for the $7,000 in lieu of fighting it out in court and the risk of losing everything. Each share was worth a meager $1,750.

Harrington and Green bantered back and forth for years after the recovery. Green claimed Harrington only located the safe because Green had marked the spot. Harrington denied his charges. Wreck diver code states that if a diver is actively working to recover an artifact, has made his intention to recover it clear and has gone so far as removing the item from the wreck and marking it with a lift bag (or in this case a buoy), then that diver retains ownership of the artifact. Apparently, wreck diver code did not exist in 1856.

Green was unable to ever dive again. However, he wrote a book entitled *Diving With and Without Armor: Containing The Submarine Exploits of J B Green* which is very much worth reading. He also made the rounds on the speaking circuit until the crowds and the funds dried up. The celebrity Green thought he deserved did not last long as the American Civil War was about to break out. People no longer cared about pompous submarine divers. There were many more important matters than his ego. Broke, washed up, living with his mother and unable to dive, he took a lethal dose of arsenic.

Rediscovery

There were several other salvage attempts after Green died. None were successful. At one time, the wreck site was very well known. However, over time, it was forgotten.

In 1984, Mike Fletcher — accomplished commercial diver of TV series *Sea Hunters* fame — rediscovered the location. This is where I will stop as an entire book could be written about the drama that ensued after Fletcher brought up some artifacts. The wreck was even closed to divers for a while. There are endless legal briefs in courts in both the United States and Canada that you can read for the details. In summary, the American wreck is owned by Canada.

Diving the Atlantic

Although some charters include the *Atlantic* on their list of wrecks she is not dived very often. She is one of the most famous shipwrecks in all the Great Lakes. Although she can be reached at a depth of 135 feet, deep diving experience is required as she rests in 160 feet of water and is a very challenging dive. Visibility on the wreck can vary from poor to well over 60 feet. More often than not, it is extremely dark. She is completely encrusted in zebra mussels.

At one time, her cargo holds could be explored with the occasional artifact being seen. However, she has deteriorated badly and is remarkably different from when rediscovered in the 1980s. Her upper decks, middle decks and cabins have all collapsed, as has her pilothouse. The wreck is heavily silted. If you will be diving this site it is imperative to visit the partially intact paddle wheels, arches and her mast which still stands proud after 166 years on the bottom.

4 Barge F

A tech diver lights the intact helm

Discovery

This unidentified barge was originally discovered by Garry Kozak during his search for the *Dean Richmond* in the 1980s. She was found again on 1 August 2001 by Jim Herbert, Kevin Magee and David VanZandt. The wreck lies at a depth of 145 feet (44 meters) in a north to south orientation with the bow to the north.

Dive

Beginning at the extreme stern of the barge (which rises about six to eight feet above the bottom) is the completely intact wheel. With the exception of a few missing handles, the eight-spoke wheel is undamaged. It is spectacular. The exposed steering gear is visible behind the wheel complete with an early chain-wrapped drum steering mechanism. Chains are visible entering under the deck through holes in the wood. These chains would have turned the tiller. At one time the rudder was partially visible, angled hard to port. The stern is beautifully rounded and is a great spot for underwater photography.

Ship

Official Number Unknown
Type Schooner barge
Built Unknown
Dimensions 135′ x 30′ (41 x 9 m)
Tonnage Unknown
Power Unknown
Builder Unknown
Owner Unknown
Previous Names Unknown
Date of Loss Unknown
Cause Unknown
Lives Lost Unknown
Location GPS 42 30.11 -79 42.00

Dive details

- **Max Depth** 145 feet (44 m)
- **Visibility** 20–60 feet (6–18 m)
- **Water Temp** low 40s°F (4–6°C)

Safety
There is sometimes a current on the surface at this wreck site. Temperatures in the high 30s to low 40s year round.

 Tech

Just forward of the wheel on the port side of the centerline is a small companionway opening in the deck. A few stairs can be seen descending below the decks. The opening is roughly three by five feet. About ten feet forward of the companionway is a larger deck opening. Adjacent to that opening is a small hand pump and a bent metal pipe.

Continuing forward is a large capstan with the following inscription: "J W Henry, Quebec and Lawrence Foundry". Joseph W Henry ran the St Lawrence Foundry in Montreal, Quebec, Canada. An early directory, from 1851, shows the foundry operated on Champlain Street. Henry was listed as an engraver and the company as having "constantly on hand all kinds of ship and mill castings, stoves".

Moving forward of the capstan is the first of five cargo holds. Actually, it is one large continuous cargo hold, divided into five distinct sections. The second hold from aft is divided in half along the centerline by a wooden beam. This is the only hold in which this appears. However, it should be noted, the other holds do have notches along their centerline which appears to suggest they may have had the same wooden beams in place at one time. Although the cargo holds are almost completely filled with silt, early divers were able to locate coal there. The holds occupy almost two-thirds of the vessel.

Once forward of the holds, a diver will encounter a large deck winch and a windlass. A second hand pump is located in the deck space between the windlass and winch. Just forward of the windlass is a towing bitt, otherwise known as a samson post.

It is rare to see an anchor still intact on a shipwreck. It is even more unusual to see two of them. However, *Barge F* is one of those rarities, and has two complete anchors visible on her forward deck. There is a fluted anchor with a metal cross piece on the starboard deck complete with a chain. The port bow also has a fluted anchor visible adjacent to the windlass.

The deck contains remarkably intact wood planking throughout. There is no railing, however, the ship does have a gunwale about one foot high. Although the ship is mostly of wooden construction, there are a few places where metal was used to reinforce the wood — along the gunwale and the cargo hold frames for instance.

Some have described the vessel as "schooner-shaped" given her graceful curves and overall shape as a sailing vessel. This is a good description of her appearance, although there is no evidence of masts or mast post holes. However, she does have a few turnbuckles on her starboard side. This suggests the possibility of the vessel having had masts at one point in time. It is possible she started out as a schooner, schooner-barge or other sailing vessel, but was later converted to a barge, thereby having her masts removed but keeping the turnbuckles.

Hopefully one day her true identity will be revealed. Even if not, Barge F is still a very photogenic wreck. Even though she rests below the recommended recreational depth limit, given the typical good visibility here the wreck can be seen from the 130 foot (40 meter) mark. At a mere 137 feet (42 meters) long, the entire barge can be seen in one dive.

> The highlight of this dive is the intact helm standing stoically on the stern

5 Cracker

A tech diver prepares to film the superstructure of one of Lake Erie's deepest wrecks

Construction

This unidentified wreck is a three-masted wooden scow measuring just under 120 feet in length with a beam of 20 feet. A very early construction (most likely built prior to the US Civil War), the vessel exhibits the trademark design of a scow — square ends. According to wreck diving lore, she was named "Cracker" by the divers who located her because the only thing they had to eat on the boat was a box of crackers.

Dive

One of the deepest dives in Lake Erie, the *Cracker* sits proudly upright at 190 feet below the surface. She must have gone down slowly as her stern cabin is intact. Typically, these cabins would have been blown off by escaping air when sinking. A few things to remember if diving this wreck: there is heavy silting, heavy zebra mussel encrustation and there is a risk of entanglement from fishing nets if not vigilant.

   Starting at the stern, although there is no wheel in place, the framework for one stands on the deck and the rudder post protrudes through. There

Ship

Official Number Unknown
Type Scow
Built Unknown
Dimensions 118′ (36 m)
Tonnage Unknown
Power Sail — three masts
Builder Unknown
Owner Unknown
Previous Names Unknown
Date of Loss Unknown
Cause Unknown
Lives Lost Unknown
Location GPS 42 33.485 -79 51.649

Dive details

- **Max Depth** 190 feet (58 m)
- **Visibility** 30 feet (9 m)
- **Water Temp** low 40s°F (4–6°C)

Safety

Large fishing net amidships which drapes the port side all the way to the bow. Penetration of the cabin not recommended due to high silting.

 Trimix

is a debate as to whether this vessel had a wheel or was steered by a tiller, given her early design. Regardless of her steering mechanism her rudder can still be seen above the mudline turned hard to port.

Her large aft cabin is still in place and is mostly intact with the exception of a missing plank or two. There is a chimney pipe extruding from the roof. The cabin has windows on the port and starboard sides but none on the forward wall. There is a companionway leading into the cabin on the port side. The cabin is mostly silted out.

Forward of the cabin there is a pump on the deck. Continuing forward, there are three cargo holds all heavily silted to the deck level. It is unknown what she was carrying when she foundered.

The bow offers plenty to see including a windlass. There is a scrolled figurehead still in place with some other carvings and designs in the woodwork above. There is a notch for the bowsprit which has fallen to the bottom of the lake off the bow.

There are two anchors off the bow — one partially buried but still attached to its chain — hanging from the port hawsepipe. The second is entangled in a heavily draped fishing net stuck off the port bow.

The ship's foremast (amazingly with top mast still attached) can be seen off the port side.

If you plan on diving this wreck you must beware the fishing nets draped on it. There is a one with two floats attached off the port stern. This floats motionless in the water. There is another net draped over the port side amidships that leads to the windlass.

> There is much to explore around the intact aft cabin

6 Dunkirk Schooner

The spectacular starboard aft view. Note the low profile cabin and intact tiller.

Discovery and controversy

In early fall 2004, I was visiting my dear friend, admiralty attorney Peter Hess,[1] in Wilmington, Delaware for a couple days. After a late night meeting and a driving tour of Wilmington we stopped at his office as he wanted to show me his brass Mark V diving helmet and some dishes he recovered from the Italian luxury liner *Andrea Doria*.

I had an early morning flight back to Cleveland, Ohio, and since the clock was pushing midnight already we decided to stay up the last few hours of my trip. After getting thoroughly destroyed in the game show *Jeopardy!* (he was the best player I have ever known), I will never forget when he looked at me and said, "Watch this!" He played some underwater footage of one of the most spectacular shipwrecks I had ever seen.

Even though the water was dark, visibility was fantastic. I was astonished to see a fully intact two-masted schooner resting upright in the lake I

---

1   Peter Hess (1959–2012) was best known for working with Gary Gentile in fighting the National Oceanic and Atmospheric Administration (NOAA) for the right to dive the USS *Monitor*. He was a famed lawyer always campaigning for divers' rights.

Ship

Official Number Unknown
Type Two masted schooner
Built Unknown
Dimensions 79' x 18'6" x 8'
 (24 x 5.6 x 2.4 m)
Tonnage Unknown
Power Sail
Builder Unknown
Owner Unknown
Previous Names Unknown
Date of Loss Unknown
Cause Unknown
Lives Lost Unknown
Location GPS 42 34.612 -79 35.002

Dive details

- **Max Depth** 170 feet (52 m)
- **Visibility** 60 feet+ (18 m+)
- **Water Temp** upper 30s to low 40s°F (3–7°C)

Safety
Bottom temperatures barely exceed 40 year round. Very dark.

★★ Tech

dived every weekend. As the video cruised over the low railing, I paused the footage when an anchor resting on the port side deck appeared. I could not believe what my eyes were seeing. I told him I had never seen that wreck before. He said no one had. I stared at Peter in disbelief and said, "What wreck is that?" He just smiled and said, "An old one".

That wreck would become known as the *Dunkirk Schooner* due to its proximity to Dunkirk, New York. She was also known as *Schooner G* and would quickly become the most controversial shipwreck in Lake Erie since the *Atlantic* made headlines in the 1980s and 1990s (see *3 Atlantic*). This unknown schooner has garnered more headlines, hate and animosity than any other wreck in recent memory. Almost everything written about it has been skewed by the respective authors' personal views. If you read some of the articles, you can sense the bias. I have attempted to separate fact from fiction, remove the bias and slant and present the matter straightforward.

In 2003 Richard Kullberg bought five sets of numbers (wreck marks) from Garry Kozak — one of which was for an intact schooner. He formed a company named Northeast Research LLC (NER) in 2004 to spearhead an ambitious project.

A 3D model view of the *Dunkirk Schooner* (PAT CLYNE)

Due to her early construction and some features found on very few wrecks in the Great Lakes, it was initially speculated that this one may have been the elusive *Griffon*. For those who are not aware, the *Griffon* is the long lost and long sought ship built in 1679 by French explorer Robert de La Salle. The *Griffon* is the most hunted shipwreck in the Great Lakes. New claims of her discovery have been made repeatedly over the 340 years she has been missing. However, each claim has proved false.

After further analysis the shipwreck was determined not to be the *Griffon*. Although she proved not to be the most famous shipwreck in Great Lakes history, the *Dunkirk Schooner* was of very early construction, and was deemed a rare wreck. She had low railings and a scrolled figurehead. One of the most unusual features on the wreck was the presence of stern gallery windows. Windows like these were only previously seen on the wrecks of the *Hamilton* and the *Scourge* — both War of 1812 ships and both constructed prior to 1808. A very remarkable discovery.

Research work

In order to attempt to identify the wreck, NER first needed to obtain a permit to do the necessary salvage. On 6 August 2004, they filed in federal court seeking title to the *Dunkirk Schooner* under the maritime law of finds, or at a minimum, a salvage award. They also sought an injunction against any rivals diving or conducting salvage operations. Additionally, NER asked for arrest papers.

The court granted the arrest and also asked for any parties with an interest in the *Dunkirk Schooner* to come forward. In September 2004, the State of New York responded and asserted the wreck was the "sole and exclusive property of the State pursuant to the Abandoned Shipwreck Act". NER's counter argument was that the wreck was not abandoned. NER believed they had discovered the long lost *Caledonia*.

On 4 June 2008, the State of New York granted a permit allowing NER to excavate the *Dunkirk Schooner* through 30 August 2008. The license was later extended for one month to the end of September. They were authorized to take measurements, collect artifacts, and excavate cargo. It addressed the always taboo subject of human remains on wrecks. The permit stated that if they were recovered NER must contact the State of New York to discuss removal and analysis.

This pocket watch was one of many items recovered inside the cabin (PAT CLYNE)

After the permit was secured, salvage and recovery operations commenced. Divers with extensive deep diving and technical diving experience were used to take measurements of the wreck. They also photographed her thoroughly. Core samples were taken from both cargo holds. Analysis of the forward hold revealed wheat and barley. The after hold contained hickory nuts.

View of a compass exposed on the interior of the wreck. Note the encrustation of zebra mussels (GARY GENTILE)

The divers used a hydraulic dredge with a small mesh-lined filter box which screened the discharge to excavate the holds. They recovered some amazing artifacts including two compasses, pottery, watches, lamps, crockery, period furniture, brass buttons and various coins ranging from 1797–1834. Some of the more remarkable finds allude to women being present on board when the ship foundered. Fragments of a Holy Bible and a publication about young women Christian missionaries were discovered along with ornate combs and feminine jewelry.

During the excavations human remains were inadvertently recovered by the dredge. About a dozen small bone fragments (a mixture of fish and human) were captured in the screen box. A forensic analysis was later completed at the Armed Forces Genetic Identification Laboratory. The results revealed the presence of a female victim.

On 21 October 2008, the State of New York alleged violations of the excavation permit due to the recovery of human remains without notice to the State, continued diving after the permit ended, and the removal of planks from the cabin roof in September 2008. There had been two divers excavating the interior of the wreck, but they discontinued operations at the end of August due to an incoming storm. They did not return for two weeks. When they did they found that the deckhouse had been torn apart.

They gathered up the boards and stacked them on the deck. They steadfastly maintain they were not responsible for the destruction.

The obvious question is, who did it? The answer will probably never be revealed. Although some person or people must know. Most of the serious wreck diving community and the general public blamed NER for the damage, while NER placed the blame on divers who were "unauthorized intruders". It was widely acknowledged that even though the wreck was placed under arrest by NER, other divers still dived it. There are even some photos available online which show divers on the wreck actually holding the arrest papers.

NER ultimately motioned for an injunction against other divers to stop them diving the wreck. That motion was denied.

On 4 March 2009, the National Park Service authorized the placing of the *Dunkirk Schooner* on the National Register of Historic Places. On 20 March that year the State of New York followed with inclusion on the State's Register of Historic Places.

The United States magistrate judge stated "clear and convincing evidence in the record establishes an inference of abandonment". NER located a descendant of Rufus Reed, owner of the *General Wayne*, who asserted her ownership. The court stated that there was no evidence of a salvage attempt after she foundered and no known attempt by any family member to locate the wreck in the 170 years she has been on the bottom and awarded title to the State of New York.

NER appealed, however, the Appeals Court upheld the initial ruling that the State of New York had title to the *Dunkirk Schooner* because the vessel was abandoned by definition of the Abandoned Shipwreck Act 1987.

Given her size and tonnage, the following ships are the ones that top a short list of possibilities. I have researched each of them thoroughly and will present their histories. Hopefully, someone will discover a clue that has been overlooked, a document that has yet to be discovered, or some identifier on the wreck that has gone unnoticed.

Dayton

The *Dayton* was a two-masted schooner of similar tonnage, but her length was less than that of the *Dunkirk Schooner*. She was constructed in 1835 in Grand Island, New York. She was built specifically for the mundane task of hauling goods from port to port. The September 21, 1835 edition of the *Buffalo Commercial Advertiser* printed the following about the *Dayton's* construction: "very burdensome and calculated for the trade of the small harbors of our Lakes".

The history of her early years is somewhat lacking as not much was printed after her launch, in 1835. However, in early March 1846, the *Dayton* was damaged in a storm on Lake Erie. In addition to flooding, she lost her bowsprit and her spars. She was later repaired. She was lost later that year in a large November gale. Sixteen bodies washed ashore between Erie, Pennsylvania and Buffalo, New York. No remnants of the ship were ever discovered. The loss of the *Dayton* is still a mystery. She has never been located.

General Harrison

Some have speculated that *Schooner G* could be the long lost *General Harrison*. Their measurements are nearly identical. The *General Harrison* was constructed in 1835 in St Clair, Michigan. According to her enrollment, she was a two-masted schooner and did have a scroll head — just like the *Dunkirk Schooner*.

The *General Harrison* had an interesting history. In a late December storm in 1841, she went ashore at Racine, Wisconsin. She was released. Seven years later she was caught in another late December storm while on the route from Cleveland to Buffalo. Her entire deck cargo of flour and seed was washed overboard. She made it to Buffalo Harbor with over three feet of water in her holds. Her cargo was ruined. However, she was lucky to have survived the storm.

No major incidents involving the *General Harrison* were recorded over the next several years. However, her last three years were plagued with disaster. In September 1852, it was reported that she sank with a load of coal, but was later raised. In August 1853, she sailed into a notorious Lake Erie storm. Tragically, Captain Edward Follett was lost when a gust of wind blew him overboard. He left behind a wife and "several small children".

In October 1854, the *General Harrison* was heading to Buffalo with 35,000 staves (to make barrels). The schooner developed a leak approximately 20 miles outside of Barcelona Creek, New York. As the leak worsened, all the men loaded into the yawl boat with the intention of leaving. As they were preparing to launch the lifeboat the schooner *Roscoe* was passing the stricken ship. Approximately 15 miles outside of Erie, Pennsylvania, the *Roscoe* took on the *General Harrison*'s entire crew and brought them to shore. The ship was lost and has yet to be found. There was one report she went ashore at Erie and broke up. However, I could not verify that claim.

Pennsylvania

The two-masted schooner *Pennsylvania* was built in Lyme, New York in 1836. According to her first enrollment, her length and depth are nearly identical to that of the *Dunkirk Schooner*. Her beam was slightly larger. Her enrollment lists her as having a plain head, not a scroll head. Very little is known about her early years, but in October 1844 she did founder approximately two miles off Point Abino, Ontario. The storm in which she was lost was described by local newspapers as a "great gale" and a "tremendous hurricane". The *Pennsylvania*'s crew of ten and entire cargo of flour and whiskey were lost. The wreck has yet to be located.

One of the signs counting against the possibility that the *Dunkirk Schooner* is that ship is that according to newspapers of the period the *Pennsylvania* had a white eagle painted on her transom.

South America

The *South America* was a two-masted schooner built in Vermillion, Ohio. The 116 gross ton vessel was launched on 24 July 1841. She had a very short career on the Great Lakes — barely lasting two years. She departed Buffalo for Toledo in October 1843 and was never seen again. Her captain and crew of six were lost. No bodies were ever recovered.

Caledonia (later General Wayne)

One of the most intriguing possibilities, in terms of American history, is the *Caledonia*. Built in 1799 on the River Rouge, in what is now known as Windsor, Ontario, the *Caledonia* was constructed for the British North West Trading Company. The schooner would work the fur trade between Fort Erie and Mackinac, Michigan.

At the breakout of the War of 1812, as the Upper Great Lakes were controlled by the British, the *Caledonia* was converted to a British troopship. She was changed from a schooner to a brig and outfitted with cannons. The *Caledonia* was used by the British in the attack on Fort Michilimackinac in 1812.

In 1813, a group of American commandos took the British by surprise and captured the *Caledonia* on the Niagara River. She was converted to an American warship and later used in, and helped win, the Battle of Lake Erie, led by Commodore Perry.

The *Caledonia* later sailed to the Detroit River where American troops invaded Southern Ontario and recaptured the city of Detroit. She also took a group of settlers and founded Fort Dearborn which later became the city of Chicago. Quite a remarkable history for a 56 foot long ship.

After her storied career, the *Caledonia* was recommended for sale. The United States government stated she was "unseaworthy from natural decay". In March 1816, she was purchased by Rufus Reed and John Dickson of Erie, Pennsylvania. The 12 March 1816 edition of the *Buffalo Gazette & Niagara Intelligencer* reported "the old *Caledonia* is rebuilding". She was to undergo several significant structural changes including being re-rigged from a brig to a schooner. She was also supposed to be shortened.

Upon her purchase she was renamed *General Wayne*. She was freshly enrolled at Presque Isle, Pennsylvania on 11 September 1816. However, no changes were noted from her previous enrollment as the *Caledonia*.

The last mention of the *General Wayne* is in the 14 August 1818 edition of the *Detroit Gazette*. She had departed Detroit en route to Buffalo. She was listed as being lost at the end of the season. The *General Wayne* never appeared in another periodical. There were never any details printed about her loss — she truly disappeared.

If the *Caledonia*'s history was not impressive enough — having fought on both sides of the War of 1812, helped win the Battle of Lake Erie, and been quintessential in the founding of one of the world's great cities — some state that the *General Wayne* even played a decisive role in the Underground Railroad, helping escaping slaves to freedom. The *General Wayne*'s owners were both active in the abolitionist movement. They represented the end of the Railroad — shipping escapees across Lake Erie to southern Ontario. According to *The Dunkirk Schooner Shipwreck Archaeological Site Assessment* by James Sinclair (2009) both the Dickson Tavern and the Reed Mansion:

> "Feature extensive labyrinths of underground passageways reportedly used to hide fugitive slaves prior to their final journey to freedom in Canada".

Other unknown shipwreck

It is also a possible that the *Dunkirk Schooner* is a wreck not in the aforementioned list. The State of New York argued that the *Dunkirk Schooner* was most likely a "nameless 1830s schooner that sank carrying grain". The 1830s may be a little late. Many believe this vessel was constructed sometime between the late 1790s and 1810, hence the list above.

So, what *is* the identity of the *Dunkirk Schooner*? That remains to be settled. There are a few important things to take into consideration before making a positive identification. First, the overall length of the vessel has been reported as 79 feet. This came from divers measuring from the bowsprit to the end of the stern yawl boat davit. In other words, the divers measured the length of the main deck. But a true measurement of a vessel's length is taken at the keel, not the deck. So, the length reported by divers for the *Dunkirk Schooner* is actually slightly longer than it would have been listed on any vessel enrollments. The overall length then must be a size that closely resembles the *Caledonia* and others previously mentioned.

Another thing to consider is that there is nothing to prove that this wreck is not the *Caledonia* or *General Wayne*. I often get asked why the court did not identify the wreck as the *Caledonia*. The follow up statement is usually "If the court did not identify the wreck as the *Caledonia*, than it must not be the *Caledonia*". It is true the court did not identify the *Dunkirk Schooner* as the *Caledonia*. However, it is very important to remember that it was not in the court's purview to do so. The court was asked to establish ownership, not identity. These are two entirely separate issues.

Plans for the wreck

Northeast Research's goal was to raise the entire wreck intact and place it on display along the Buffalo waterfront. She would be the main attraction of a maritime museum. The wreck would be fully submerged in cold water inside an acrylic tank. How would this be possible? Well, the wreck is less than 80 feet long. There are tanks at aquariums that are larger. Imagine standing in a museum and coming face to face with an intact shipwreck built in the 1790s. That would be a worldwide attraction and would no doubt leave an indelible mark on those who came to see her.

Given her pristine condition and very old construction, plans were made to raise the wreck and conserve her for future generations. Some have criticized the move, online and in articles, as careless, irresponsible and monetarily driven, but without any details having been publicly released. I can ensure you that the plans for conservation are much more involved than what has been insinuated in the media. Just because most (specifically those who were complaining the loudest) were not privy to the information, does not mean that there were no plans. In other words, just because an army doesn't submit its battle plans to the enemy, it doesn't mean that they don't have a strategy.

Diving the Dunkirk Schooner

Full profile view (CAPTAIN STEVE GATTO)

A dive to the *Dunkirk Schooner* is really a tale of two dives — one prior to and one subsequent to the destruction of portions of the wreck after her exact location was published. I'll detail some of the changes.

She sits upright in 170 feet of water with a slight list to starboard. Visibility is typically very good on the wreck with 50 plus feet expected. It has been known to reach upwards of 100 feet. As a diver descends to the wreck, the upright masts — which rise to a depth of 100 feet — can be seen coming into view sometimes as shallow as 60 feet. Both have crosstrees, but do not have top masts. Dual standing masts are a rare sight.

Commencing at the bow — which points west — an ornate, scrolled figurehead can be seen. This is a spot divers must visit and is unmissable for those with underwater cameras. The bowsprit is no longer attached. However, it rests off the starboard side of the stem, angling down into the bottom.

The port bow is highlighted by the presence of the port anchor laying on the foredeck. Chains still adorn the anchor and run across the railing to the stem. Aft of the stem is a windlass and a samson post.

Continuing aft, the foremast is still standing proud. Four deadeyes used to be visible nearby until they were removed. A boom rests next to the starboard railing on the main deck.

While heading to the mainmast, one will swim over a small cargo hatch. At one time this hatch was filled with silt but it has since been excavated. After that is the mainmast. There used to be three deadeyes on the railing.

Starboard bow view. The figurehead can be seen on the far right. The broken bowsprit is in the foreground. (CAPTAIN STEVE GATTO)

A unique feature of the masts is that each contains an integrated fife rail on the back side of the mast. Fife rails are typically found at the base of the mast, surrounding the mast's entire circumference. These are typically destroyed when masts break, although there are still some wrecks which have intact ones. However, it is a very interesting feature and one that I have not seen previously.

Continuing aft of the mainmast is the aft cargo opening — roughly the same size and position as the fore cargo hatch. Just beyond this is the aft cabin, which is very low profile for a cabin. The cabin was previously fully intact with the exception of a few missing planks of wood. All of the roof boards have been removed and stacked on the main deck along the exterior cabin walls.

A unique feature outside on the main deck is a pair of large hand pumps located on the forward corners of the cabin. The cabin was fairly full of silt until it was excavated. Stairs used to be seen heading down into the ship. Shelves could also be seen below decks. I write "used to" because the stairs have been ripped out and pieces of wood and furniture have been strewn about the main deck.

A large tiller (precursor to the helm or wheel) was once intact and looked as if it could still be used to steer the ship. Unfortunately, the tiller now lays on the deck having been removed from the rudder post. Sadly.

The transom is unique. This is another must stop for the underwater photographer — I highly recommend putting it on your shot list. There are two large, very rare, uniquely-placed square windows on the transom, which look directly into the cabin. You can imagine staring at the captain looking out into the lake from these windows in the early 1800s.

The intact rudder can be seen underneath the transom. Due to the schooner's list starboard, the entire underside hull on the port side is open for viewing — a very impressive sight.

The port yawl boat davit is visible at the extreme stern, as well as the uniquely shaped transom with full gallery windows and fully exposed rudder (CAPTAIN STEVE GATTO)

The *Dunkirk Schooner* is also known as *Schooner G* and the *Admiralty Wreck*

Dunkirk Schooner

One of two hand pumps located outside the cabin on the main deck (GARY GENTILE)

Left: An anchor rests on the port bow.
Right: A close up view of the gallery windows on the transom.
Note the notch on the top of the rudder. (GARY GENTILE)

The port bow of the *Dunkirk Schooner*. The foremast remains as if she is still sailing. (GARY GENTILE)

7   George J Whelan

A deep tech diver swims along the remains of the George J Whelan

Construction

The *George J Whelan* had a short life by comparison to other vessels of her time. However, in the 20 years between her launching in 1910 and her ultimate foundering in 1930 she left quite a legacy on the lakes. Having endured five name changes, three sinkings, registries in four different countries on both sides of the Atlantic, service in World War I, and a plethora of other incidents, the *Whelan* has an impactful, if not tragic, story. Unfortunately, most of that story would be lost to history after she foundered … until now.

The *Whelan* began her illustrious and ill-fated career as the *Erwin L Fisher* in June 1910. The *Fisher* was built for the Argo Steamship Company of Cleveland by the Toledo Shipbuilding Company. She was named after the Argo's managing director.

Small in stature compared to other steel hulled vessels constructed during that time, the *Fisher* measured only 220 feet in length, with a beam of 40 feet and a cargo hold depth of just over 15 feet. She was known as what was commonly referred to as a "canaler". These types of vessel, although small, were preferred for the coal and lumber trades as they were able to pass the tight canals and wind through the twisted tributaries. This

Ship

Official Number 207617
Type Converted sand-sucker
Built 1910
Dimensions 237' x 40' x 17'
 (72 x 12 x 5 m)
Tonnage 1,293 gross tons
Power Steam engine
Builder Toledo Shipbuilding Company, Toledo, OH
Owner Kelley Island Lime & Transportation Company, Sandusky, OH
Previous Names *Erwin L Fisher, Bayersher, Port De Caen, Claremont*
Date of Loss 29 July 1930
Cause Load shift
Lives Lost 15
Location GPS 42 25.555 -79 44.986

Dive details

- **Max Depth** 150 feet (45 m)
- **Visibility** 30–70 feet (9–20 m)
- **Water Temp** low 40s°F (4–6°C)

Safety

There are multiple entry points inside the wreck. However, the passages are tight especially with stages, camera equipment, etc.

 Tech

enabled them to get to the smaller Great Lakes ports more easily than ships of larger dimensions.

Whatever she lacked in size, she more than made up for in power. Her massive triple expansion engine was powered by two scotch boilers made by Lake Erie Boiler Works in Buffalo. When at full tilt she could put out over an astounding 800 horsepower, with her single screw getting a remarkable 125 revolutions per minute. Very impressive for a relatively small ship.

Fisher wrecked

On 14 July 1910, the *Erwin L Fisher* completed her maiden voyage without incident after delivering a load of 2,000 tons of coal. This would be the first and one of the last times she was in the newspaper for anything other than disaster or death. Her next headline was emblazoned all in capitals in the *Buffalo Evening News* of Friday 5 May 1911:

"STEAMER FISHER SINKS IN RIVER, THREE MISSING".

What exactly transpired is still a mystery. As my dad always told me, there are three sides to every story — the first party has their version, the second has theirs, and the truth lies somewhere in the middle.

1910 launch of the *Erwin L Fisher* at the Toledo Shipbuilding Company

In service in 1911

As with the majority of maritime accidents, lawsuits were filed. It can be difficult to locate any paperwork from a case filed over a century ago. However, if you know where to look, you can be successful and I was able to locate the court rulings. After pouring through legal filings, testimony, newspaper articles, investigation reports and records, this is what I believe to be accurate regarding the tragic events of 4 May 1911 on the Detroit River. For a quick geography lesson, the western end of Lake Erie is connected to Lake Huron via the Saint Clair River, Lake Saint Clair and the Detroit River.

There are several islands and islets throughout the winding Detroit River, but it is one of the most essential rivers in the Great Lakes.

During the late evening hours, while the clock was pushing midnight, the *Fisher* was bound up the Detroit River. She had departed Lorain, Ohio en route to Port Arthur, Ontario with a load of soft coal in her holds. She also had 700 tons of steel rails destined for the railroad on her deck. There was a contingent of 15 aboard.

Meanwhile, the huge steel hulled freighter *Stephen M Clement*, under the command of Captain Henry H Townsend, was bound down the Detroit River with a cargo of iron ore. Owned by the Buffalo Steamship Company, the 5,821 gross ton ship measured 480 feet long with a beam of 52 feet — much larger than the *Fisher*. She was crewed by 23.

Sometime between 2320 hours on 4 May and shortly after midnight on 5 May (based on all the testimony, newspaper accounts and the United States Steamboat Investigation Service report), the two ships collided off Grassy Island in the Grosse Ile Channel. The *Fisher* was struck by the bow of the *Clement*, aft of the pilothouse, on the port side.

> "While the crew of the *Fisher* awakened by the shock of the collision were tumbling from their berths and running to the rail to jump overboard, their vessel turned on her side and went to the bottom".

The *Fisher* went down fast. After being struck she rolled over on her port side. The gaping hole allowed the vessel to fill rapidly, giving the 15 on board very few moments in which their fates were decided. To say all hell broke loose is an understatement. Three jumped overboard into the dark waters in their night clothing. Nine others reportedly leapt onto the bow of the *Clement* before the two vessels separated. Chief Steward Lewis Sudgen, the assistant steward (Lewis' wife Kate) and the chief engineer were missing and presumed dead.

Clarence Pashaw was one of the three crewmen who threw themselves into the river in a desperate attempt to save himself. Here is his account of his harrowing escape in his own words:

> "I was lying awake in my bunk about midnight when suddenly there was a giant crash and I was thrown out on the floor. Water was pouring through the portholes and down the ladder ways. I rushed for the stairs, only to find tons of water coming down like a flood. How I reached

the deck I do not know; but I fought my way up somehow. The sound of the escaping steam, the crashing off of plates and the rolling of the 700 tons of steel rails made speech impossible. Then we began to go down. I guess I saved myself by jumping out into the river as far as I could and swimming until I was picked up by the *Clement*'s yawlboat".

The *Clement* immediately anchored and her crew began an intense search for anyone in the water. They could not locate anyone besides Pashaw and the two others who had jumped in and were picked up by the yawl boat. The *Clement* remained on site until 1500 hours, at which time she continued her journey to Buffalo to offload her cargo of ore at the Buffalo & Susquehanna Coal and Iron Company. Even though her bow sustained damage in the collision, it was not serious enough to impede her journey. She was taken to dry dock and repaired. Upon his arrival in Buffalo, Captain Townsend refused to comment.

On 6 May, after the Argo Steamship Company abandoned the *Fisher* to the underwriters as a total loss, and while that same insurance company solicited bids for her salvage, the body of the chief engineer was recovered. The husband and wife stewards were still missing. They were presumed to have been trapped and to have drowned in either the dining room or their berth when the ship went down.

Two days later, the Great Lakes Towing Company started the salvage operation. They won the auction with the lowest bid: $39,000 or 40 percent of the value of the property recovered. However, before they could begin the process of raising the ship, they first needed to recover the 700 tons of steel rails. It took nearly a month for Captain Alex Cunning, wrecking master, and his three ships — tugs *Favorite* and *Roth*, and lighter *White* — to raise all the steel.

On 11 June, they were finally able to raise the port side of the *Fisher* four feet off the bottom. Five days later they successfully righted the ship on an even keel. The work moved slowly:

> "It was feared it might prove difficult to overcome the suction of the muddy bottom on the side of the steamer. The method employed is very similar to the way lumbermen load logs on a wagon — five jack screws adding their lifting power while the wrecker *Favorite* pulls on chains passed about the hull".

Once she was righted they were finally able to explore some of the interior of the ship. Sadly, they recovered the bodies of Lewis and Kate Sudgen.

In July they were able to start constructing a cofferdam. Finally, on 6 August at approximately 0600 hours, three entire months after she foundered, they were able to start the pumps. It took less than five hours to raise the ship. Once she was refloated she was taken to the shallows where the coal was offloaded. After that she was moved to the Great Lakes Engine Works Yard in Encorse, Michigan to await her new fate. Captain Cunning said it was one of the most difficult jobs he had ever done.

What makes this tragedy astonishing is that the collision happened on a calm and clear night in an 800 foot wide channel with no other boat traffic in sight! There was definite negligence, but on whose part? The courts would ultimately decide, but the case would take years to come to a final resolution.

Legal aftermath

There were three separate investigations and/or hearings concerning the collision between the *Clement* and the *Fisher*. Of course, all three had different conclusions concerning liability. The first ruling came down on 28 October 1911 from the United States Steamboat Inspection Service. According to their report Captain Henry H Townsend of the *Clement* violated Pilot Rules I, II, III and V of the Great Lakes, as well as Rule 17 of the laws relating to navigation. Blame was placed squarely on the *Clement*. Captain Townsend's license was suspended for only 39 days. I guess 13 days suspension for each of the three lives lost was deemed sufficient. This decision was appealed against and overturned by the Supervisory Steamboat Inspector.

While the United States Steamboat Inspection Service investigation was ongoing, a lawsuit was filed in the United States District Court for the Eastern District of Michigan. In February 1913, after a lengthy trial in which the lawyers for the *Clement* contested liability, Judge Arthur J Tuttle ruled in favor of the *Clement*. He believed the *Fisher* bore sole responsibility for the collision and ruled that the *Clement* was blameless. It is interesting that the first two rulings, by two separate bodies, came to opposite conclusions even though they had the same set of facts.

The lawyers representing the *Fisher* immediately appealed to the United States Circuit Court of Appeals for the Sixth District. Twenty-seven months later, in May 1915, Judge Warrington reversed the decision of the lower court and placed blame equally on both vessels. The testimony was relatively simplistic. The *Fisher* stated that the *Clement* ported and was crowding the *Fisher* prior to impact. The *Clement* maintained that the *Fisher* ported, then starboarded immediately in front of the *Clement*'s bow, which caused the *Clement* to plow into the *Fisher*.

In fact, the only thing the two parties could completely concur about was that there had been an accident. Although a port to port passing was initially agreed upon, each vessel stated that they sounded their horn a certain number of times for different reasons, which the other vessel could not figure out. The *Clement* did admit to deeming the passing dangerous, but proceeded anyway. That may have been the deciding factor when the court ruled they were partly to blame.

Judge Worthington summarized this nicely:

> "The testimony of the two navigating crews resulted as usual in two distinct and opposed theories, and of course, if each ship had been navigating along the course her crew described, the admitted accident could never have taken place".

The judge continued:

> "The navigating conditions were entirely consistent with a safe passage of these ships. Admittedly, the night was clear and the lights of the ships and on the adjacent shores were alike easily discerned. There were no intervening boats to disconcert the eye and no wind or noise to disturb the hearing. The channel with abundant water was 800 feet in width and the sailing line was straight and in the center of the channel between the points at which the boats were perspectively sighted".

In other words, the collision should not have happened. He concluded:

> "If we rightly interpret the evidence touching the *Fisher*'s movements, there is no perceivable way to account for the disaster at all except through a movement of the *Clement* in material degree to the eastward" (port, towards the *Fisher*).

One would rightly think he is blaming the *Clement* for the collision. However, he noted that the lookout of the *Fisher* was not in place during the passing. He wrote "our study of the records has convinced us that both steamships were in fault".

And with a stroke of the pen, the case was decided. The first investigation revealed a ruling in favor of the *Fisher*. The second was in favor of the *Clement*. The final one placed blame on both vessels. My dad was right — the truth does lie somewhere in the middle.

Repaired

The 1911 salvage of the *Fisher* (LOWER LAKES MARINE HISTORICAL SOCIETY)

Because the Argo Steamship Company abandoned the vessel at the time of the sinking, the underwriters were forced to pay out the $105,000 insurance policy even though the ship was raised. At the end of September 1911, as a way to recoup some of their losses the underwriters offered the wreck for sale. All bids were initially rejected as too low. However, at the beginning of October, they sold the wreck to the Toledo Shipbuilding Company for $42,750 — a full $10,000 more than the highest bid received during the first week. They had bought themselves a severely damaged ship in desperate need of repair. After the sale was complete she was towed to Toledo to begin restoration for her new life on the Inland Seas.

While still undergoing work in Toledo, the Argo Steamship Company re-purchased the vessel. On 20 January 1913, she was back in service, repaired and complete with some modifications to enable her to operate in saltwater. Eight months after the *Fisher* was wrecked, she successfully completed a trip to Kingston, Ontario, hauling a load of grain. She made her return trip to Duluth, Minnesota with a cargo of ore.

The rest of the 1912 shipping season was quiet. She was berthed in Cleveland over the winter. However, in March 1913, while she was being readied for the upcoming season, they discovered that she was grounded in port due to the bottom of the lake shifting during the winter and encapsulating the ship. She was dug out.

War service

The *Fisher* continued to operate in the Great Lakes until March 1916 when she was purchased by the Lake Transportation Company. She was immediately sold to the Bay Steamship Company of the United Kingdom. Because she was a canaler and able to ply tight, windy waterways, she was acquired to augment the war effort. Unlike most Great Lakes ships, which were purchased for use along the United States East Coast during World War I, the *Erwin L Fisher* was destined for Europe. She departed in September and arrived in Montreal to meet her new owner. Her name was changed to *Bayersher* and she was sent to the European theater.

Records are scarce as she was in the service of a foreign government and it was over a century ago. What is known is that, while operating in the English Channel in 1918, the ship struck a mine. Knowing she would sink, the captain immediately turned her towards the English coast in a desperate and valiant attempt to save the ship and the lives of those on board. She sank for the second time before she made it. However, she sank in the shallows. She was later raised, repaired, and returned to service. She continued to serve the British until 1921, three years after the war ended.

Becoming the George J Whelan

Later in 1921, the British sold the *Bayersher* to the French government, who in turn, renamed her *Port De Caen*. A year later she was sold to Canadian parties who renamed her *Bayersher*. The following year she was acquired by Captain Misener for use by the Dominion Sugar Company. Her name changed once again. This time she was named *Claremont*, which was the estate name for the sugar company's president, Ralph Gilchrist. She had two groundings while named *Claremont* — the first in 1926 on Lake Ontario and the second in 1929 when she was driven ashore in a storm on the Saint Lawrence River. Repairs for the 1929 grounding cost $8,760.

She sailed under the name *Claremont* for seven seasons until 1930, when she was sold to the Kelley Island Lime and Transportation Company for $60,000. Upon her sale she was renamed *George J Whelan*. She was purchased specifically in order to transport sand and was therefore converted from a bulk freighter to a sand barge, otherwise known as a sand-sucker. This change required many modifications, all of which raised the center of gravity and made the ship top heavy. An A-frame and cargo boom towering 90 feet above the deck was constructed. In addition, sand boxes were added to her holds. Each box measured over 80 feet in length, with a width of just under 31 feet and was over twelve feet high. The major issue with the boxes was that they were constructed eight and a half feet above the bottom of the hull. Due to this, they protruded three feet above the deck.

Her troubles began immediately upon acquisition and were a foreshadowing of what was to come. On 27 July 1930, one day prior to departing on her fateful voyage (which was her first trans-lake crossing as a sand-sucker), one of the *Whelan*'s crew members, Frank Looker, was killed on the boat when he was struck by her heavy machinery.

Final journey

On 28 July, less than 24 hours after Frank Looker was tragically killed, the *Whelan* departed Sandusky, Ohio for Tonawanda, New York. She carried 21 people on board — Captain Thomas Waage, 19 crew, and the wife of the cook. She was loaded with 1,611 tons of crushed limestone — more than her capacity. Limestone was piled several feet above the sandbox hatches. This had a domino effect. As the limestone was protruding, the hatch covers would not fit and they were therefore left off. This exposed them to the possibility of water ingress if the ship encountered a storm or heavy seas. Secondly, due to the excessive cargo, the giant, heavy unloading boom could not be secured in its proper place. If this was a movie, this is where the haunting music would start playing as a herald of what would happen next, and the director would zoom in on the mounds of limestone and insecure boom.

Although the weather was clear when they departed, a storm was brewing. It finally manifested itself at 0030 hours on 29 July. The *Whelan* began to list ten degrees to starboard due to the wind. In order to correct it, Captain Waage ordered a change of course. It worked for a moment, then the ship listed to port. At that time, the Captain left the wheelhouse to confer with his chief engineer, Arthur Walters. Waage wanted Walters to use the water ballast to right the ship. The captain was never seen again. First Mate Irving Ohlemacher was in the wheelhouse. As the ship continued to list heavily to port, the order was given for the crew to come topside. The *Whelan* could not call for assistance as she did not have a radio.

The list was reportedly so extreme that they could not launch a single lifeboat. It happened so fast that it was almost impossible for the entire crew to make it on deck before she flipped over. The time was approximately 0130 hours. Luckily, she remained afloat even though she was upside down. Eight lucky men clung to the overturned hull in desperation. However, they did not know that they only had 30 minutes before the ship would disappear. When she finally succumbed to Lake Erie and slipped beneath the surface, one of the eight men also disappeared. They were down to seven and only two had lifebelts. Arthur Stamm and Eckhart Lange kept Ohlemacher alive by sandwiching him in between their preservers.

After two hours, Stamm started to swim to shore — a distance of eight miles. He later stated:

> "There was nothing to do but swim for it. I found myself in the water without a life preserver and sensing the direction of the shore, started out. I must have swam about a mile and my arms had hardly begun to tire when I sighted lights of a ship".

Stamm spotted the massive 525 foot long *Amasa Stone*. Thankfully, First Mate Robert Endleman heard the cries in the distance and notified the *Stone*'s captain, Walter H McNeill.

Captain McNeill immediately ordered the engines stopped and gave the command to lower the lifeboats. Stamm was the first picked up as he was the closest. He was later quoted as saying:

> "My cries were heard by the captain of the ship, which happened to be the *Amasa Stone*, it was short work of picking me up, although it was very dark. I sure was glad to see that ship and will always be thankful to her captain. I believe I could have made shore all right. It was only

about eight miles they tell me. However, that is chiefly a guess and I have no regrets that I did not have to make that swim".

Five others were picked up after Stamm. Literally moments before the men were rescued, while the lifeboat was in sight, Chief Engineer Arthur Walters, who was being assisted in staying afloat by Lange, told him "I'm gone!", slipped from his grasp and disappeared beneath the surface.

Captain McNeill radioed to shore that a vessel had sunk off Dunkirk, New York. He kept the *Stone* in the vicinity until daybreak while they searched for survivors. McNeill said he heard more cries in the darkness, but that the source could not be found. Imagine hearing cries in the dark distance, knowing men were in the water but you could not find them. Then imagine the silence after the cries stopped. It must have been torture.

At approximately 0515 hours, United States Coast Guard (USCG) Station Erie had a telephone call from the Officer in Charge of USCG Station Cleveland. They had received:

> "[A] radiogram from the steamer *Amasa Stone* of the Interlake Steamship Lines, that he had picked up some men 22 miles west of Dunkirk, NY on the Buffalo, NY to Erie, PA course and to send the Coast Guard at once".

According to the USCG Station Erie logs, they "immediately manned picket boats *CG-2259* and *CG-9013* and started for the scene".

The *Charles Donnelly* arrived at approximately 0530 hours to assist the *Amasa Stone* in the search. The steamer *Greater Detroit* joined them and stood by to assist. As word of the sinking spread throughout the maritime community, other vessels came to offer any assistance they could to their missing brothers.

At approximately 0630 hours, the picket boats met with the *Stone*:

> "… about 18 miles out of Erie, PA, and were informed by the Master W H McNeill that the sand-sucker *George J Whelan* had foundered at a position 22 miles west of Dunkirk and about two miles inside the Buffalo and Erie steamer lane".

The picket boats "cruised in the vicinity of the disaster" until 0913 hours at which time *CG-9013* discovered they were getting low on fuel. *CG-2259* continued the search while *CG-9013* headed back to Station Erie. When *CG-9013* returned to the scene, they brought extra gasoline for both picket boats and for the seized *Robert B* who assisted in the search for survivors.

At 1215 hours, Station Erie launched power surfboat *2176* to meet the *Stone*. The six survivors were transferred to *2176*, and were taken to the sand dock in Erie where they were given warm, dry clothes and aid.

At 1252 hours, the USCG Officer in Charge met the survivors and delivered a message to Ohlemacher that he must wait at the dock until Steam Boat Inspectors Nolan and Todd arrived.

At 1401 hours, *CG-106* arrived at Station Erie with the body of Thomas Pierce, the fireman.

The USCG launched a massive search operation. In addition another nine coast guard vessels searched in vain for survivors. For nearly a week, dozens of boats, including fishing vessels and other privately owned craft, plied the waters looking for any sign of the wreck and her crew. No other bodies or survivors were found.

A clipping from the *Sandusky Star Journal*, July 29, 1930

Investigation

Captain William P Nolan and Captain James M Todd were the investigators assigned to the *Whelan* disaster. The survivors, however, were reluctant to speak. Thus, very few details were known at that time. First Mate Ohlemacher (the only officer rescued) issued the following statement from the dock:

> "It is the custom of the sea not to make comment on what has happened until we have conferred with the owners of the vessel. We shall pursue the same course. There is nothing further to tell regarding what happened until official statements are made or until an investigation is made".

It was then that Ohlemacher turned towards Lake Erie and said of McNeill,

> "I'll always remember what you did for us, Captain. You certainly saved us. I and the rest of these men owe you a lot and we certainly will not forget it".

Ohlemacher never spoke of it again.

The surviving crew was interviewed and they related some egregious details. One said that "the high waves were tossed up after the wind suddenly shifted". They also stated that as the limestone absorbed more and more water, the sand-sucker began to list to port. They were called on deck and were told to sit on the starboard side. There was sufficient time to lower the lifeboats, however, the order was never given. They believed that they survived because they were thrown clear of the vessel when she suddenly flipped. It was reported in the 30 July 1930 edition of the *Sandusky Register* that there was a severe electrical storm the night of the sinking, but the survivors maintained that the seas were calm. Their best guess was that the limestone cargo shifted, which caused the list.

Although the exact cause of the sinking remains uncertain there are a couple of theories. What is known, is that Ohlemacher did not want to speak ill of Captain Waage. He thanked the captain for everything he did to save the crew despite the other survivors' thoughts regarding the non-launching of lifeboats. Ohlemacher was loyal to his captain. Whether he thought differently in private is not known, but what he said publicly was that his captain did no wrong.

Some of the crew speculated that the ship had a leak, which mixed with the limestone cargo and caused the stone to move more easily. This caused the load to shift, which in turn made the ship list. This seems to agree with the United States Steamboat Inspection Service's view. In their report they pointed to the fact that Captain Waage did not verify the stern drain boxes were closed. If they were indeed left open, this would have allowed massive quantities of water to pour into the cargo holds and mix with the limestone over many hours. This most certainly could account for the shifting load.

Inspectors Todd and Nolan also held the captain responsible for overloading the vessel. The holds were overflowing with limestone. This stopped the hatch covers from being installed. It also prevented the cargo boom from being stored correctly. It is possible that a heavy boom swinging wildly to one side could cause the ship to list.

Despite the lack of clear evidence as to the exact cause of the sinking, the newspapers declared Captain Waage innocent. Large, bold font emblazoned the following headline in the 29 July 1930 edition of the *Sandusky Star Journal*:

"INSPECTORS EXONERATE CAPT. WAAGE".

The court of public opinion was in session and they refused to blame a dead captain who left behind a newborn child and a widow. Inspector Todd issued the following statement:

> "There is not one lot of evidence to show that Captain Waage did not act judiciously in an attempt to save his crew. He had a reputation for good seamanship and fearlessness".

This despite the views of the surviving crew.

Regardless of how she foundered, of the 21 on board at the time of sinking, only six survived.

The 15 who perished, according to officials at Kelley Island Lime and Transportation Company, were: Thomas Waage, Captain; M J Emling, Second Mate; Arthur A Walters, Chief Engineer; Carl A Blechele, First Assistant Engineer; William Neucheler, Second Assistant Engineer; Ned DuMurs, Derrick Operator; Charles Godfrey, Steward; Mrs Charles Godfrey, Assistant Steward; Ralph Weis, Watchman; Thomas Pearce, Fireman; Ed Donner, Fireman; John Stanley, Fireman; Harry Brooks, Oiler; W P Longnecker, Deckhand; and Arthur Zeck, Watchman.

Diving the George J Whelan

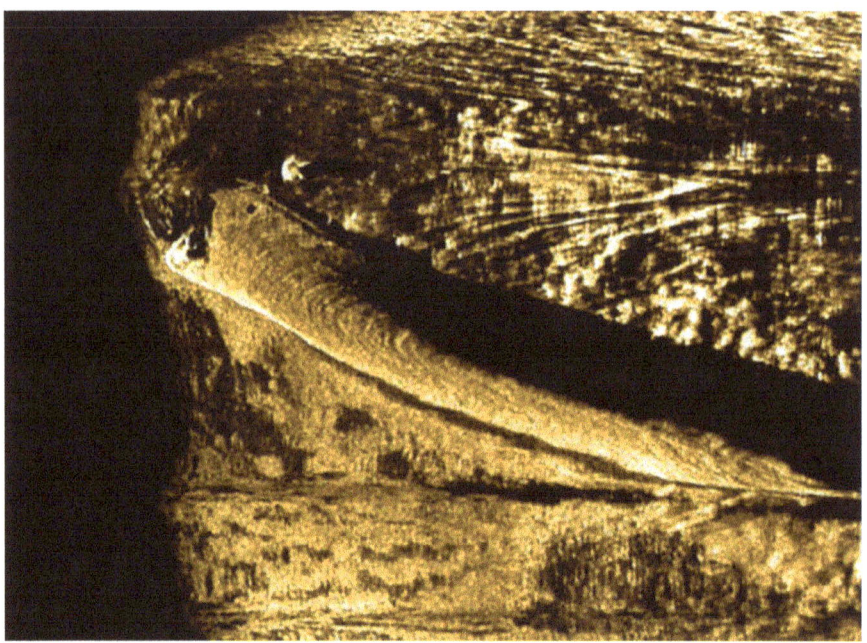

Side scan image of the *Whelan* (GARRY KOZAK)

The wreck of the *Whelan* had been forgotten to history, but not to divers and wreck hunters. Those of us who dived the lake often knew she was out there waiting to be discovered. For years Garry Kozak and Captain Jim Herbert had gotten together to search portions of the lake for her remains. Locating a shipwreck is never easy and it is never in the first place you look. Just as Indiana Jones says, "X never marks the spot!"

In late October 2005, after conducting additional archival research and pouring through newspaper accounts, they selected a 32 square mile section of water where they believed the *Whelan* might be found. Armed with the latest side scan sonar technology, they swept that area in only ten hours! Seventy-five years after she foundered with the loss of 15 lives, they discovered the *Whelan* resting on her port side in 145 feet of water.

The first divers down noticed her pristine condition. Artifacts were strewn about and the portholes were all open. Kerosene lamps, fire extinguishers, and porcelain light fixtures could be seen.

She rests on her port side but is almost completely turned over. Visibility is usually great on this wreck and can reach upwards of 60 feet. One of the first things a diver will notice is the large four-bladed propeller with the giant rudder behind. Also at the stern, there is an opening large enough for a diver to swim inside, where a hallway and room can be seen. Inside there are several artifacts including bunk beds and a lamp. There is other debris in the area. There are many open portholes along the stern.

Approximately 30 feet off the ship is a partially buried yawl boat. This is one of the lifeboats which the surviving crew stated they could have launched, if the order had ever come. This is a fantastic photo opportunity.

> The open portholes along the *Whelan*'s hull allow divers to peer inside and glimpse many artifacts

Continuing forward one will encounter the cargo deck and the open hatches. There is a lot of debris outside the stern superstructure. Further on towards the bow, divers can really see the expanse of this wreck. Although the front portion of the bow is partially buried, there are more open portholes. Peering through into the interior of the ship is akin to looking back in time. More remnants of everyday life at sea are evident, including lanterns.

The *Whelan* is truly a remarkable shipwreck and, even though her deepest point is 145 feet, she can be dived comfortably at 125–130 feet. It's also a photogenic wreck, so if you have a camera or video system I recommend taking it with you.

Two tech divers head towards the *Whelan*'s stern (WARREN LO)

8 John J Boland Jr

Stern view of the overturned ship. One of the first things a diver sees while descending is the giant four-bladed propeller and rudder.

Construction and naming

The *John J Boland Jr* was built by Swan, Hunter and Wigham Richardson Ltd at their Neptune Yard in Wallsend-on-Tyne, England. The shipyard was located just downriver from Newcastle. A common misconception is that the constructor of the *Boland* was "Wallsend-on-Tyne", that however is a place, not a shipbuilder.

Most publications have listed the *John J Boland Jr* as "ex-Tyneville" — as the common consensus is that she was launched as the *Tyneville*, but whose name later changed to the *John J Boland Jr* sometime later. Even Lloyds Register lists her as "ex-Tyneville". However, my research with the Swan Hunter archives revealed she was named in the yard list as *John J Boland Jr*. This suggests her name was changed before she was completed. The *Boland* was launched on 23 March 1928 and made her way across the Atlantic without any issues.

The 1,939 gross ton steel-hulled, steam propeller was a bulk freighter. In the Great Lakes these ships were also known as "canalers" as they needed to ply and traverse the tight, small locks and canals in the region. The *Boland* was operated out of Saint Catherines, Ontario, Canada, by Sarnia

Ship

Official Number 149467
Type Bulk freighter
Built 1928
Dimensions 252'9" x 43'4" x 17'8" (77 x 13 x 5.3 m)
Tonnage 1,939 gross tons
Power Triple expansion steam engine
Builder Swan, Hunter & Wigham Richardson, United Kingdom
Owner Sarnia Steamships Ltd, St Catherines, ON, Canada
Previous Names Tyneville
Date of Loss 5 October 1932
Cause Foundered in storm
Lives Lost 4
Location GPS 42 22.794 -79 43.929

Dive details

- **Max Depth** 137 feet (42 m)
- **Visibility** 60 feet+ (18 m+)
- **Water Temp** low 30s to mid 40s°F (0–7°C)

Safety

There is sometimes a current both at the surface and on the wreck. Aft Cabin and forward superstructure can be penetrated. Beware of disorientation as the wreck is almost upside down. May not be able to fit with inside doorways with CCR and stages. Access to the engine room is possible but very silty and narrow passages along the route. Reduced visibility inside the wreck.

★ Deep

Steamships. As a bulk freighter she was built to haul grain, coal, and other big loads between ports in Lake Erie, Lake Ontario, and various other places along the Saint Lawrence River.

Final journey

On the night of 4 October 1932, the *Boland* took on a freight of coal in Erie, Pennsylvania. However, she was overloaded to the point where an extra 400 tons of her cargo flowed out of her hatches and covered the deck. She was laden to the point she reached her maximum draft. There were no tarps or hatch covers fastened to the load. This was the common practice of the time and would unfortunately soon prove to be her demise.

The *Boland* was bound for Hamilton, Ontario, by way of the Welland Canal — which connects Lake Erie to Lake Ontario with a series of locks. This allows ships to safely bypass Niagara Falls on the journey. This was considered a routine route for her.

The weather forecast issued on 4 October 1932 was as follows:

> "Rain tonight and Wednesday; colder Wednesday along the Lakes; low 70, high 76".

The following morning, it was:

"Cloudy and colder tonight with occasional light rain; high 56, low 52".

That predicted "light rain" began a period of 48 straight hours of rain from a huge storm no one had anticipated. Being the shallowest of the Great Lakes, Erie is infamous for sudden, violent storms, treacherous waters and swelling seas, which as history proves have been an obstacle for many a mariner.

The *Boland* left the docks of Erie at approximately 0300 hours with her load of coal, 19 crew (including the wife of the first mate), and $3,000 in Canadian cash for the payroll. Yes, the money went down with the ship for those who are wondering.

As the *Boland* made her way northeast across Lake Erie and away from the protection of the American shoreline, the water seemed to come to life. She "ran into dense fog and heavy seas a few hours out of Erie". However, the *Boland* was "moving steadily on its course". On this day, the waves grew with every passing minute. Soon the deck was awash. The coal, piled high and exposed, was washed overboard as the hatches began to fill with water.

The brave crew made haste and battled the wind and waves while trying to secure tarps over open passageways and hatches. Little did they know that they had only about four minutes before the *Boland* would disappear into the dark abyss below.

Captain Edward C Hawman, at the helm, recognized his ship was in peril and tried a desperate maneuver to bring her around and head towards the safety of the American shore. However, as fate would have it, the rudder would not respond. The *Boland* was "spinning around in the storm".

Huge waves crashed over the deck and the ship quickly developed a list which made it impossible to launch the lifeboats. One of the survivors was quoted as saying:

"We all dashed across the deck and tried to cut loose the boat on the other side. This one made the water all right".

Captain Hawman, First Mate M Smith, his wife, and Chief Engineer William Byers all rode in a lifeboat together. It is interesting to note that the three highest ranking officers made it into a lifeboat. They were "tossed helplessly" for four hours before they were able to land in Barcelona, New York. They made their way to a local farmhouse owned by Joseph Bissennette for shelter.

Captain Hawman stated that the *Boland* "sank in less than four minutes". The local newspaper notified United States Coast Guard 161, under the command of Boatswain Sivert E Hunshedt, of the disaster. He ordered his crew be ready and immediately contacted the United States Coast Guard at Buffalo and Erie.

Eleven others jumped or fell overboard when the *Boland* listed heavily to one side. They swam with all haste towards any wreckage they could find. Some managed to find planks floating in the water. One of the survivors set the scene of the dramatic final moments of the *Boland*:

> "I saw her turn upside down. The keel was sticking straight up in the air. She settled stern first right after that and I saw a column of steam shoot up into the air".

Miraculously, these eleven individuals "narrowly escaped death" and floated for an astounding 32 hours before landing on Bournes Beach, New York. They were "barely able to stand" from exposure. The survivors were taken to local farmhouses and were given food and clothes.

There is an amazing side note to this story about the *Boland*'s oiler, Brownie Sadler. He was one of those who drifted for many hours. It was the third time Sadler had to be rescued from a shipwreck. He was on the SS *Norwood* in 1914 on Lake Michigan, and again in 1918 on the *Chesapeake Bay* he was saved from the sinking SS *Texana*. In an interview after the *Boland* sank, Sadler stated he would be back on a crew in the spring, once the lake shipping commenced, as "it is all a part of life at sea".

Sadly, Sidney Brooks (32 year-old oiler), Harry Jones (41 year-old oiler), George Geary (22 year-old fireman), and Jean MacIntyre (24 year-old second cook) perished. MacIntyre's body was recovered on 5 October, approximately five miles off Barcelona by the fishing tug *Betty & Jean*. Her body was entangled in the fishing net the boat was hauling aboard. She was wearing her life jacket. The bodies of Brooks, Jones, and Geary were never recovered. It is believed they were trapped below decks in the engine room at the time of the sinking.

Aftermath

The loss of the *John J Boland Jr* received an abundance of attention in the news. In addition to the stories published in Canada, the sinking was reported in newspapers from coast to coast in the United States — from Massachusetts to California — and from as far north as North Dakota to as far south as Texas. In all, 16 states' presses covered it.

The loss of the *Boland* came 20 years after the tragic sinking of the *Titanic*. Even though the great number of lives lost aboard the *Titanic* can be attributed to there being insufficient lifeboats to save every passenger and crewman aboard (among other reasons), customs 20 years later in the Great Lakes still included reckless practices like overloading cargo holds and covering decks with freight. All in order to make a buck.

During this time period in the Lakes there were no rules or regulations governing the loading or stowing of cargo. In fact, vessels were filled with cargo until the captain said otherwise, even if that meant overflowing the holds. If he wanted to completely fill the holds and then add even more tonnage so that the load spilled out onto the deck in order to meet the maximum draft of the vessel, the captain could do so. As the master of the ship, he was given complete deference as to how cargo was loaded. The common opinion of the day was that it was the captain's ship and he could do what he wanted with her.

These same issues were the focus of a hearing regarding the loss of the *Boland* held on 22 October 1932, in Montreal, Quebec, Canada, before the Dominion Wreck Commissioner. The tribunal was ordered by the Canadian Government agency — Department of Marine. Captain Hawman was tried on the charge of Evasion of Precautionary/Safety Measures due to alleged customs prevalent on the Great Lakes. The Dominion Wreck Commissioner, L A Demers, suspended Captain Hawman until 1 January 1934 and demoted him to first mate. This was ordered even though Demers found there was "no willful neglect". Additionally, the *Boland*'s first mate was warned while the second mate was exonerated.

Commissioner Demers continued by strongly suggesting the need to "abolish pernicious customs said to exist on inland seas". Specifically, he was focused on indifferent stowing, trimming of cargoes, and carrying of deck loads. The loss of the *Boland* combined with Demers' recommendations helped change practices on the Great Lakes. Sometimes life is born from tragedy. There is no doubt these changes helped save countless lives and ships on the Great Lakes in the years that followed.

Diving the John J Boland Jr

Anyone who knows me well knows that I have an intimate relationship with the wreck of the *Boland*. I have published accounts of my near-fatal dive there. Still, the *Boland* is one of my favorite wrecks in the Great Lakes. She rests at 137 feet beneath the surface, on her starboard side. Rising 40 feet off the bottom, she is quite impressive. The highlight of the dive is the giant four-bladed propeller, which is typically the first thing divers see as the chains for the dive buoy are attached there. The rudder and rudder post are intact and massive. Just aft of the stern cabin is a small stern deck which is partially buried in the mud.

The stern cabin is interesting. There are several decks, deck winches, and ladders to explore. There are two portholes and two open doors leading into the cabin. Access to the galley can be had by traversing some corridors and making some turns through these open doors. I suggest running a line. I would also suggest studying the ship blueprints to know how to get there and back out of the wreck. There is some refrigeration equipment to see along with some china plates. The engine room is open, but difficult to get to and even more difficult to penetrate, as it is a tight squeeze and easily silts out. However, divers can see the raised skylights to the engine room, even though they are partially buried.

Proceeding forward from the aft section, one will encounter a pair of lifeboat davits — a haunting reminder of the tragic sinking in 1932. If your buoyancy is perfect you can swim underneath them, just above the bottom of the lake, on your way forward to the cargo holds. The five holds are penetrable and, as one would expect, cavernous.

The forward superstructure can be explored through an inviting door at the aft end. This leads to two rooms through a wood-paneled hallway. This is the area where some of the lower officers' quarters were located. Unfortunately, the wheelhouse is buried. Be wary of penetration here as I know several divers who have been temporarily trapped inside the forward cabin by falling debris. The silt is very thick and visibility is quickly reduced to zero. Due to the list and position of the wreck, the ceiling, floor, starboard and port sides are the opposite of how they would appear if the ship was upright. Divers be forewarned of this.

> A fantastic shipwreck with an abundance of opportunities for penetration and photography

The wreck of the *Boland* is an impressive dive and there is a lot to see on the outside of the wreck without ever venturing inside. However, should you choose to explore the interior, use caution.

As a side note, in August 2007 I was contacted by Lewis Brooks regarding the *Boland*'s deceased sailors. He wrote:

> "Do you have a list of the deceased from the *John J Boland* shipwreck? My father lost his life when the ship went down. I am 79 years of age and would like to know before I join him. Thanks".

His father was Joseph Sidney Brooks — one of the oilers on the ship. I asked whether he had any photos of his father, or of the *Boland*. Sadly, Lewis lost all of his possessions, including photos of his father and the ship, when his family home burned to the ground. I contacted the Swan Hunter shipbuilding archives in England to inquire about the *Boland* ship file. I was looking for any documents, photos, or deck plans of the *Boland* while she was being built. Unfortunately, I was informed that none of this survived.

Through some exhausting research I was able to locate a set of deck plans, as well as some historical newspapers and archive photos. I also sent a list of the deceased to Lewis Brooks. I hope I was able to provide him with some much needed closure for this tragedy.

A collection of newspaper clippings covering the sinking

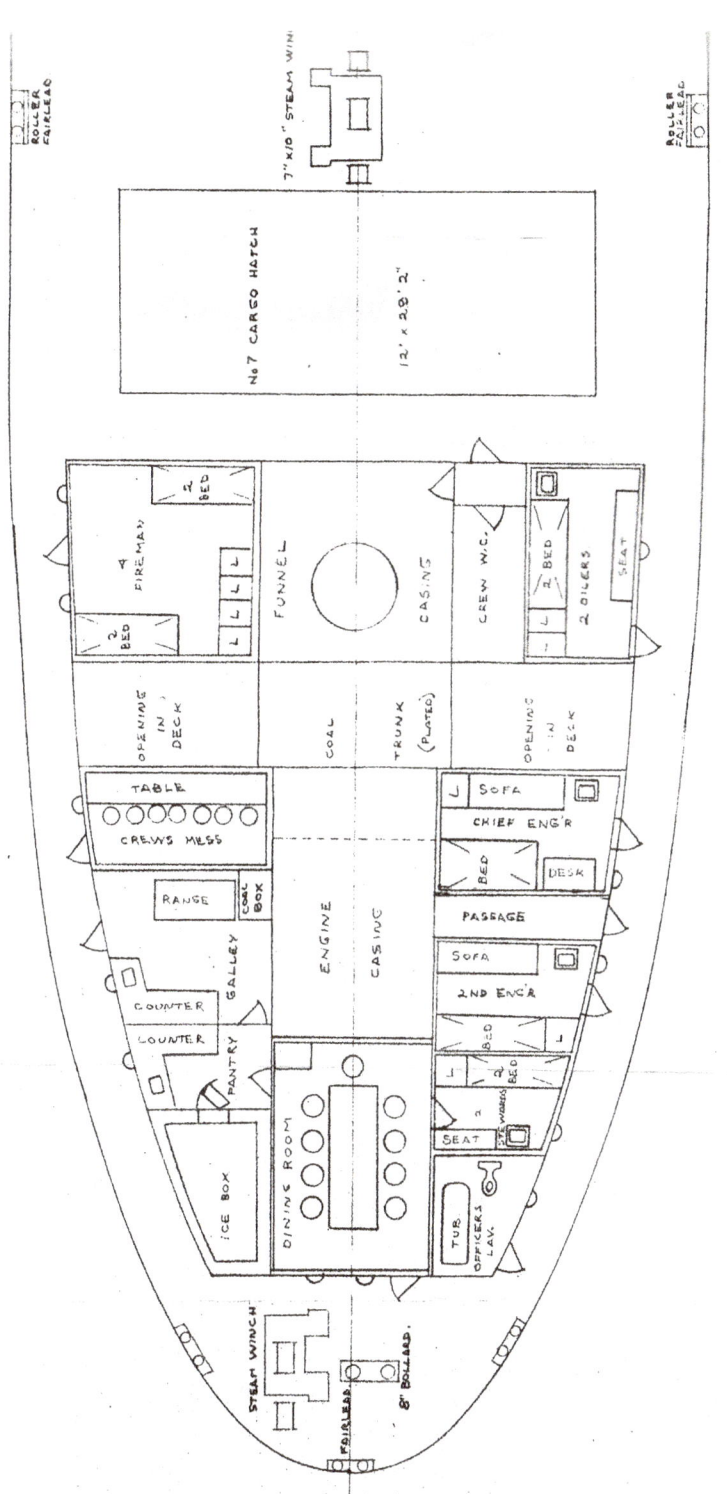

Deck plan

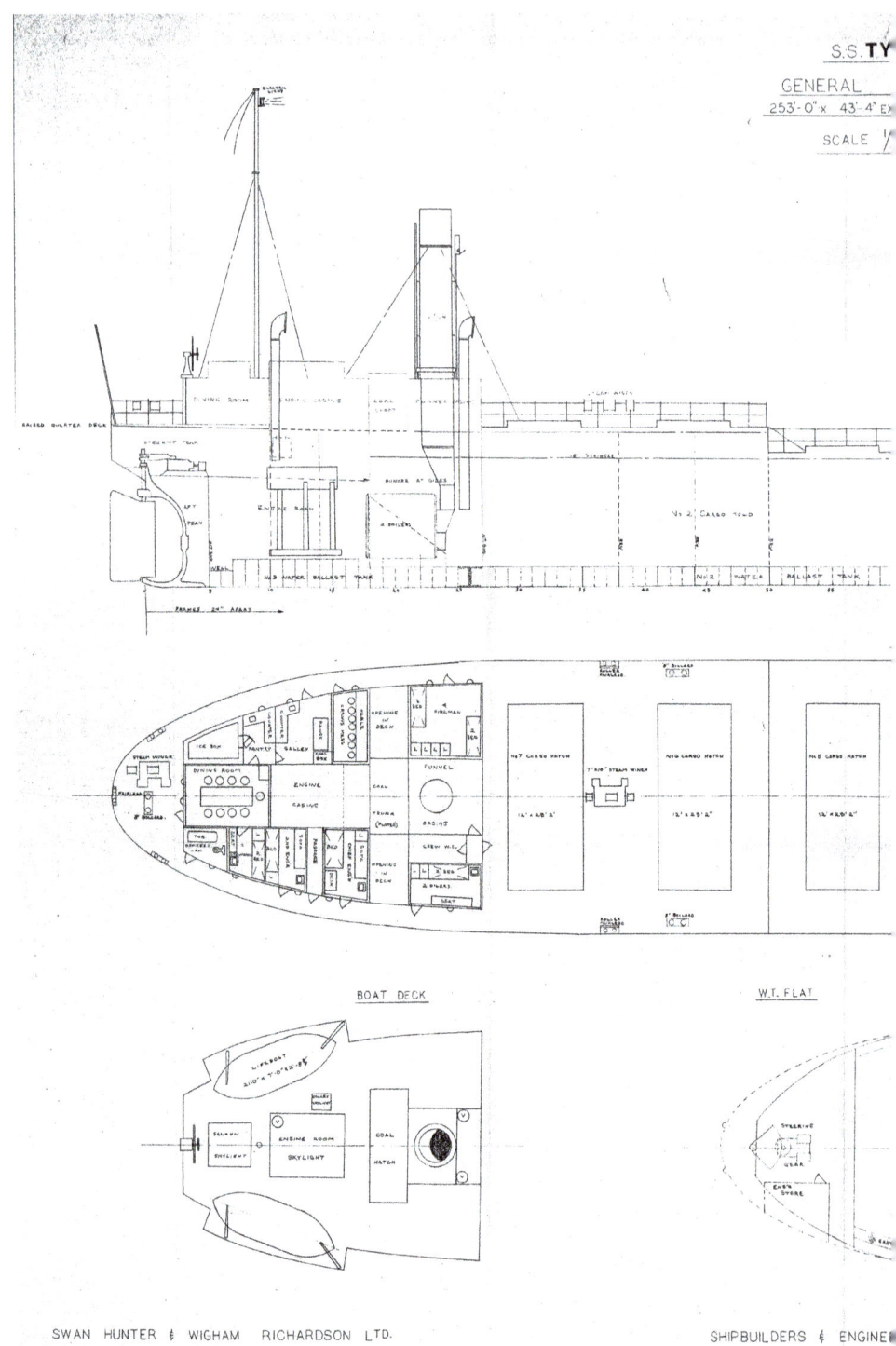

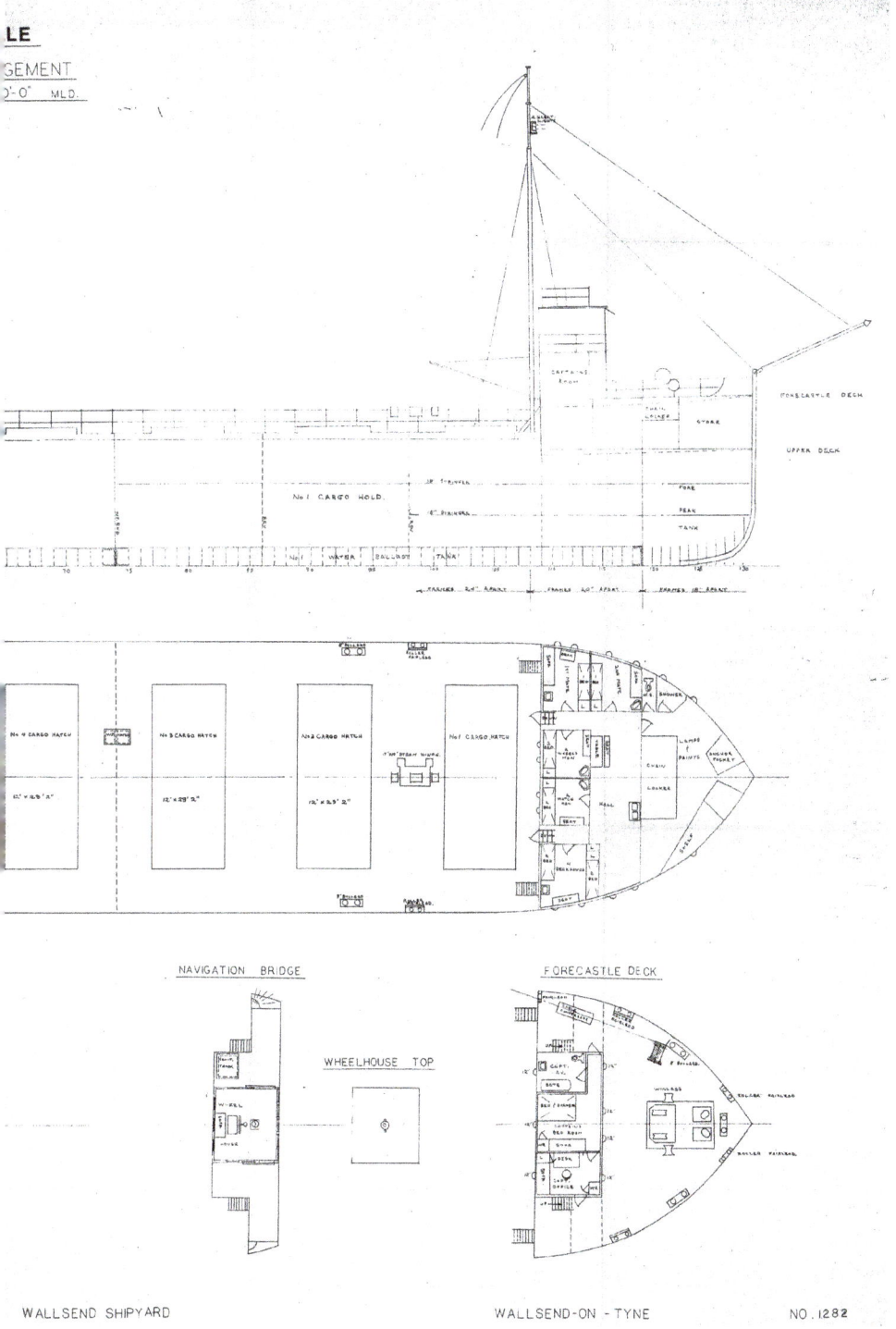

9 Junction 20

The "Junction 20" wreck is very seldom dived. This unidentified wooden sailing vessel rests at 162 feet. She is badly broken up. The location is in Canadian waters to the northeast of Long Point, Ontario — a long spit of land that juts out 20 miles, half way across the lake. This spit has been the cause of many scores of shipwrecks with countless lives lost. Ships were forced ashore in wicked storms when the fierce winds and currents drove them onto the land. Long Point was almost unavoidable at night in the days before navigational equipment and radar. In fact, the first lighthouse was not erected until 1830. It only lasted twelve years. A second lighthouse was constructed in 1843 and was in place until the early 20th century.

Diving the Junction 20

Named for her proximity to an underwater gas well, the wreck was discovered (and heavily damaged) when a gas pipe was laid and installed in the 1990s.

Although almost completely buried by silt, her foremast can be seen — it has fallen forward and rests off the port bow next to the bowsprit. There are also signs of transom damage. Heavy and costly

Ship
Official Number Unknown
Type Scow
Built Unknown
Dimensions Unknown
Tonnage Unknown
Power Sail
Builder Unknown
Owner Unknown
Previous Names Unknown
Date of Loss Unknown
Cause Unknown
Lives Lost Unknown
Location GPS 42 35.402 -79 58.378

Dive details
- **Max Depth** 162 feet (49 m)
- **Visibility** 20–50 feet (6–15 m)
- **Water Temp** upper 30s to low 40s°F (3–7°C)

Safety
Ship is heavily broken. Almost completely buried by silt.

★★ Tech

excavation would be needed in order to begin to identify this wreck.

On May 8, 2015, the fishing tug *Iron Fish* was trawling approximately six miles off Long Point when its net snagged an object. They hauled up an intact mast and towed it to port. It measured 23 meters tall and reportedly weighed nearly 15 tons. The mast was estimated to be 200 years old and from the unidentified Junction 20. The mast was an immediate tourist attraction. However, due to the cost of preservation, it was decided to tow the mast back to the lake and put it back where it was taken from.

> Despite being heavily broken and almost completely covered, the bowsprit remains off the port bow

10 Mast Hoop

This unidentified wooden schooner lies at a depth of 170 feet on the bottom of Lake Erie. Because of her final resting place — due north of the tip of Long Point, Ontario, Canada (see page 100) — there can be a strong current here.

The stern rises only a few feet above the mud line, while her starboard side is completely buried in the muck. The intact bowsprit proudly extends from the bow as if she is still sailing. A long section of her mast measuring approximately 30 feet lies off the port side.

The wreck is known as the "Mast Hoop" because of those which are found on the wreck. Although today some mast hoops are made of plastic and other materials, the ones on this wreck were made of wood, as they were for most sailing vessels in the 19th century. Some hoops were made of iron. The mast hoop simply connected the sail to the mast.

She is also known as "Bob Powell's Wreck" in some publications as Bob is the commercial fisherman who first reported her location.

Ship
Official Number Unknown
Type Schooner
Built Unknown, likely mid-late 1800s
Dimensions 145′ (44 m)
Tonnage Unknown
Power Sail
Builder Unknown
Owner Unknown
Previous Names Unknown
Date of Loss Unknown
Cause Unknown
Lives Lost Unknown
Location GPS 42 33.418 -79 59.524

Dive details
- **Max Depth** 170 feet (52 m)
- **Visibility** 30 feet (9 m)
- **Water Temp** low 40s°F (4–6°C)

Safety
Very strong currents. Wreck partially buried in the sand.

★★ Tech

The bowsprit still extends from the bow

11 Oneida

A painting of the Oneida from 1869

Research

Different publications — newsletters, books and others — sometimes disagree when it comes to certain details. Oversight, mistakes and negligence may play a role. Oftentimes, this information is then copied from one source to another. After many years and multiple publications these errors can become ingrained in the story. Maritime history is no exception.

When researching this ship, one of the issues was that there were multiple vessels with the same name in the same time period. Steamers, schooners, propellers, barges, tugs and a yacht were all called *Oneida*. It has been reported elsewhere that the *Oneida* on the bottom of Lake Erie, at this specific location, was built in various years ranging from the early 1840s, to 1855, 1859, and even as late as the early 1890s. However, based on my research, this *Oneida* was built in 1846 and sank in 1852, and what appears here is the correct history for this wreck.

Ship

Official Number None assigned
Type Hogging arched propeller
Built 1846
Dimensions 138'3" x 24'1" x 11'
 (42 x 7 x 3.3 m)
Tonnage 428 gross tons
Power Steam engine
Builder B B Jones, Cleveland, OH
Owner O A Knight, Cleveland, OH
Previous Names N/A
Date of Loss 11 November 1852
Cause Storm
Lives Lost 17–25
Location GPS 42 13.966 -79 51.583

Dive details

- **Max Depth** 160 feet (49 m)
- **Visibility** 40–50 feet (12–15 m)
- **Water Temp** upper 30s to low 40s°F (3–7°C)

Safety
Stern is draped by fishing net. Thick silt on the deck.

★★ Tech

Construction

The first appearance of the *Oneida* in any publication I could locate was in the 16 June 1846 edition of the *Buffalo Express*. Under the heading "Lake Intelligence" the periodical announced her recent launching. Built for Pease & Allen & Burke by Benjamin B Jones in Cleveland, Ohio at a cost of $26,000, the initially 345 gross tons package freighter measured over 138 feet in length, with a beam of 24 feet. The graceful steamer had two decks, two masts and hogging arches along each side. Hogging arches were used in early steamship construction in the Great Lakes for structural support prior to the advent of better hull designs. She was built to haul cargo on the line from Buffalo to Chicago.

Even though she was only two years old, the *Oneida* underwent significant renovations in April 1848. The owners had one of her masts removed and added a third deck. This ultimately changed her capacity from 345 to 428 gross tons.

Incidents

Shortly after her renovations she was involved in two rescues less than a week apart. On 2 July 1848, the schooner *Cadet* with a load of lumber was hit by the steamer *New Orleans* approximately six miles outside Buffalo Harbor. The *Oneida*, on her return route from Buffalo to Chicago, spotted the *Cadet* in "sinking condition" and towed her into port prior to foundering. Four days later, she towed the capsized schooner *Gallinipper* into Buffalo.

In October 1848, the *Oneida* barely averted disaster. She was off Kelleys Island in Lake Erie's Western Basin when she collided with the steamer *Arrow*. The *Oneida* sustained unknown damage described as "considerable". The collision was so violent that the *Arrow*'s wheelhouse was carried away. Both vessels were lucky not to have foundered. They were later repaired.

In each of the next four years the *Oneida* was involved in some sort of incident in the Great Lakes. In August 1849, her propeller shaft broke in transit while in Lake Michigan. She was towed into Chicago by the steamer *Ohio*. On 11 October 1850, she collided with the steamer *St Louis* off Vermilion, Ohio with little damage. Once again, she was lucky to escape a collision without sinking.

In December 1851, less than a year before the *Oneida* would make her final voyage, she encountered a wicked winter storm characterized as "one of the most fearful gales ever experienced on Lake Erie". This storm chased the steamer *Atlantic* (see *3 Atlantic*) back to the safety of the harbor in Cleveland, wrecked the steamer *Mayflower*, and caused the *Oneida* to lose her rudder. Miraculously, she beached just below the pier at Fairport Harbor, Ohio. She was pulled off and her rudder was repaired. Unfortunately, five months later in May 1852, she would collide with that very same pier and damage her hull. But her luck of escaping disaster was about to change — her last voyage was less than six months away.

Loss

On 11 November 1852, the *Oneida* departed Cleveland with 3,500 barrels of flour — 1,100 of these were placed on her main deck as her cargo holds were full. She stopped at the railroad depot and received additional freight — unknown quantities of ham and beef divided into 14 pound measurements called tiercins (an Old English unit of measurement). The ship was heavily loaded. In actuality, she was overloaded. There was somewhere between 17 and 25 crew aboard when she departed on her journey. She was heading out into a storm.

Sometime that evening in the middle of the violent gale, the steamer *Keystone State* passed an overturned vessel — bottom up. The *Keystone*

State's captain deemed it to be the *Oneida,* based on her hull paint.

A local newspaper, the *Kingston News*, ran a 27 Nov 1852 article entitled "The Lost & The Living" succinctly summarizing the deadly tempest:

> "The fearful November storm which swept over the chain of Western Lakes brought sorrow and desolation to many a domestic hearth. The total number of lives lost will probably exceed 60, most of them suddenly engulfed in surging waters".

It was catastrophic — 30 vessels ashore or grounded, 55 damaged, and seven lost, including the *Oneida*.

It is not known how the ship sank. Yes, the storm played a key role. However, what specifically caused the ship to founder (a leak, or overloading perhaps) may never be revealed. What is certain is that the *Oneida* never made it to Buffalo. With the exception of her overturned hull, she was never seen again. There were no survivors. No bodies were ever recovered.

Although the ship was never sighted again, remnants of her loss could be found everywhere. For nearly nine miles along the Dunkirk, New York shoreline the beach was "strewn with portions of the wreck and cargo". Her promenade deck, doors, windows and a portion of her forward bulwark bearing her name were found. Two yawl boats emblazoned *"Oneida"* came to rest on the beach. Interestingly, one of the boats had the ship's papers, books, and due bills lashed inside. Someone from the crew had enough time to grab the papers, place them in the boat, and tie them down for safekeeping. Whether that person made it off the ship or into the yawl is not known. Since no bodies ever washed ashore, if that person did get into the lifeboat, he did not stay in it.

It must have been a wicked storm. Barrels of flour from the *Oneida* washed up in Erie, Pennsylvania — a distance of nearly 50 miles from Dunkirk.

It is unknown how many were on board, or who, thus estimates range from 17–25 lost. However, the families of the officers paid a heavy price. Captain William Rich left behind a wife and four small children. First Mate Samuel Hulgate had a wife, a child, and two sisters. Second Mate C William also left behind a wife and child.

Diving the Oneida

There is still some speculation as to the identity of this wreck. There are three possibilities given the type of vessel and the era in which she was built: she is either the *Idaho*, *Ohio*, or *Oneida*. The *Idaho* sank in shallow water, which precludes that possibility. The wreck cannot be the *Ohio* as she suffered a massive boiler explosion. There is no evidence of a blow-up and this wreck's machinery is not damaged. The *Oneida* fits the bill: a hogging arch package freighter built in the right time period; the presence of only one mast; the overall dimensions of the wreck; and the lack of superstructure (as most of it floated off and landed on shore as previously described).

> The massive, intact, dual hogging arches stretch over 100 feet on both sides

The *Oneida* rests at a depth of 160 feet approximately six and a half miles off the Long Point Lighthouse. Visibility is not great here compared to other wrecks in the Eastern Basin, averaging 40–50 feet at depth. Although her superstructure has disappeared, the wreck offers much for the underwater explorer. The hull is intact and the deck rises approximately six feet above the bottom.

Starting at the bow, which stands ten feet above the lake bed, dual fluted anchors can be seen. This is a rare sight and a must stop for anyone diving this wreck. A windlass is located aft of the anchors on the forward deck.

Heading aft, some open hatches can be seen along the deck. A mast hole is visible, as is the broken mast resting along the starboard side. The cargo holds are open for viewing and some penetration can be done here as there are several feet of space above the silt. However, maintain proper buoyancy to avoid a complete silt-out.

The stern offers a view of the large engine, with the boiler visible below deck. A fishing net hangs off the transom. A portion of one of her propeller blades can be seen protruding from the mud.

The highlight of a dive on the *Oneida* is viewing the dual hogging arches still in place. The arches span over 100 feet along each side of the wreck. This is a great opportunity to view early steamship design up close. The arches are massive and rise to a depth of 125 feet.

12 Oxford

An inviting hatch opening on the deck

Construction

The year 1848 was a time of strife and upheaval throughout Europe. Revolutions and protests occurred in Sicily, France, Germany, Italy, Belgium, Denmark, The Netherlands and throughout the Austrian Empire. The potato famine was occurring in Ireland. Karl Marx had just published his *Communist Manifesto* in London. In the United States, the country was expanding. Gold was discovered in California. The Mexican-American War had ended with the United States acquiring Texas, California, New Mexico and parts of Utah, Colorado, Arizona and Nevada. The first telegraph link was established between New York City and Chicago. The Great Lakes were now connected to the rest of the country via wire.

    Meanwhile, in Chaumont, New York — a small village at the far eastern end of Lake Ontario — local shipbuilders were at work constructing vessels for the burgeoning economy throughout the Great Lakes region. The Copley and Mann shipyard was busy raising two boats — one of which was the wooden two-masted brig of 254 tons named *Oxford*. On 23 February 1848, it was announced that she would be ready in time to set sail "on the opening of navigation for the upper lakes".

Ship

Official Number None assigned
Type Two masted brig
Built 1842
Dimensions 114' x 24' x 9'
 (35 x 7 x 2.7 m)
Tonnage 254 gross tons
Power Sail
Builder Copley & Mann Shipyard,
 Chaumont, NY
Owner Hoag Strong & Company,
 Cleveland, OH
Previous Names N/A
Date of Loss 30 May 1856
Cause Collision
Lives Lost 5
Location GPS 42 28.855 -79 51.843

Dive details

- **Max Depth** 160 feet (49 m)
- **Visibility** 50 feet+ (15 m+)
- **Water Temp** upper 30s to low 40s°F
 (3–6°C)

Safety

Temperatures in the high 30s and low 40s year round. Very strong currents have been reported on the surface.

 Tech

Working life

Not much is known about the early years of the *Oxford*. The first mention of her in local papers after her launch was on 23 October 1851, when she went ashore at East Sister Island. East Sister is west of Pelee Island in Lake Erie's Western Basin. She was later released along with her cargo of iron ore, although she had lost $2,000 worth of it.

Two years later, on 10 November 1853, the *Oxford* was en route to Toledo, Ohio, with a load of railroad iron when she encountered a severe storm approximately 15 to 20 miles outside Oswego, New York. The captain stated "The wind was very severe during a portion of the night". So severe, in fact, that the *Oxford* was completely dis-masted. She somehow made it through the night despite losing her foremast, mainmast, bowsprit and all sails and rigging. The steamer *Bay State* was sent out the following morning, after the storm had passed and towed her into port. She would sit out the remainder of the 1853 shipping season.

On 31 May 1854, the *Oswego Times and Journal* reported on the upcoming season. They wrote:

"We noticed that the *Oxford*, which was used so roughly in a gale last fall, [is] recovering her former appearance. Both her spars were carried away — and in fact everything above deck — which are now being replaced by the finest looking 'sticks' we have ever seen".

Her 1854 season may have started well, but it would end in disaster. An early December storm swept across Lake Erie and Lake Ontario and the effects would be felt in the entire region. The schooner *R R Johnson* wrecked at Airport, Ohio. The schooner *Mansfield* wrecked at Cleveland, Ohio. The steamer *Ontario* went ashore off Nicholson Island in Lake Ontario.

In the same storm, the *Oxford* went ashore at Cape Vincent in Lake Ontario. She dragged her anchor in an attempt to stay off the rocks, but struck a pier and ultimately sank. Even though she filled with water and her cargo of corn was damaged, she fortunately sustained no major damage. Due to her position close to shore and "out of the way of the lake seas" she was able to withstand the strength of the storm without further damage and could later be raised.

Collision and sinking

On 29 May 1856, the *Oxford* was loaded with 360 tons of railroad iron destined for the upper lakes. Captain John Lee and his five crewmen were accompanied by his wife and young daughter for the journey. The weather and seas were calm. They made their way west across Lake Ontario from Oswego and into Lake Erie without incident.

Meanwhile, the propeller *Cataract*, owned by the American Transportation Company, was sailing east from Toledo with a load of flour and general provisions. Her ultimate destination was Buffalo, New York.

In the early morning hours of 30 May, the *Cataract* and the *Oxford* were both navigating the waters off Long Point, Ontario. The weather remained calm, the sky was clear, and visibility was not hampered by darkness. In fact, the lights of the *Cataract* could be seen by the *Oxford* from a distance. The *Oxford*'s light was burning, however, it was dim. Her first mate was instructed to keep her away from the oncoming ship. But he left his post and went to the forecastle to light his pipe. Upon his return from below decks, the *Cataract* struck the *Oxford*, abaft (nearer the stern than) the foremast and sliced her to the waterline.

Knowing his vessel could not survive, Captain Lee — who was on deck at the time of the collision — rushed below to his stateroom to rescue his wife and child. All three were never seen again. The *Oxford* sank in less than three minutes. The mate who abandoned his post drowned, as did one additional crewman. Three survived. The *Cataract* continued on its way to Buffalo where she arrived with only $100 worth of damage.

People in Oswego and the Lake Ontario region were appalled by the tragedy and the supposed carelessness of the *Cataract*. The 4 June 1856 edition of the *Oswego Times and Journal* called for an investigation, using some pretty strong words:

> "The recent disaster on Lake Erie, by which the brig *Oxford* was lost, should be judicially investigated, and not permitted to pass by unnoticed as was that of the reckless, and perhaps criminal, destruction of the steamer *Northerner*. Five human beings were drowned, or if the sinking was caused by want of vigilance, or recklessness, murdered, and a proper regard for the public welfare demands a thorough judicial investigation. We do not believe all of the many accidents on the Lakes are the result of accident, or even carelessness".

Shortly after the collision, US Steamboat Inspectors Charles Brown and Thomas Truman opened an investigation. Although the people of Oswego would get their wish, the outcome may have surprised them, as no blame would be placed on the *Cataract*. In the inspection report, the following ruling was made, in September 1856:

> "1st — That no blame can attach to Capt. Hunt, or to the officers of the *Cataract*, as all the testimony goes to show that they were at their posts; that Capt. Hunt, especially, was on the forward upper deck, keeping a good look out; that when he saw the red light he gave the order to port the wheel, which was in accordance with the standing rule, as passed by the Board of Supervising Inspectors and the common practice on the lakes; that when he discovered that the vessel, instead of keeping away, the tendency of which course was to place the vessel in such a position as would inevitably bring about a collision, Captain Hunt promptly and at once rang his bells to stop and back, at the same time hailing the vessel to luff, and keep her course, to which no attention was paid by those having charge of the vessel, was promptly done by the officers of the propeller *Cataract*.

2nd — That the sole cause of the collision was the order given to the mate of the *Oxford*, to keep her away, and there is no doubt in our minds but that the order was given whilst the mate was confused, he having just come up from the forecastle, where he had been to light his pipe, and that had he remained on deck and kept watch of the light, no collision would have taken place.

It is proper for us to say, that many vessels are navigated on the lakes without showing any lights, and, in many vessels that have lights, no proper care is taken of them, and that towards morning the lights are burning dim and almost out, as was the case in this instance. We cannot but condemn a practice so obviously tending to endanger the lives of all who are compelled to follow the lakes, either for profit, as passengers or otherwise, at the same time making the duties of the Licensed Pilots very arduous, frequently placing them in situations of peril, as well to life as reputation".

The public outrage seemed to be directed at the wrong ship — at least according to the official investigation. A quick light of the pipe is all it took for five people to lose their lives. I am unsure if there is any solace knowing Captain Lee's family perished together — it must have been some horrific and terrifying final minutes for the three of them, especially the little girl.

Diving the Oxford

The wreck of the *Oxford* rests in a mud bottom, down at 165 feet. Although this is the maximum depth, the wreck can be comfortably dived in the 145 feet range. Although heavily broken at the bow, the wreck offers a tremendous sampling of early Great Lakes ship construction. There is a plethora of artifacts along the entire wreck, including at least a dozen deadeyes, belaying pins and other rigging. This wreck usually has excellent visibility, so I would suggest bringing your camera.

Beginning at the bow — which was heavily damaged by the collision with the *Cataract* and faces west — the scrolled and cutwater bow can be seen in the mud. There is also a large amount of wooden debris in the area. There is an anchor off the port bow.

Just aft of the bow on the port side, the foremast is fully visible — although she points diagonally towards the surface. The foremast contains the crosstree — which some have mistakenly called a crow's nest. This feature has led some to refer to the wreck as "The Crow's Nest". The topmast is still connected.

The crosstree still attached to the foremast is sometimes mistakenly referred to as a crow's nest (CRIS KOHL)

Continuing aft, the rails are fairly intact, with deadeye and belaying pins. There is a large bilge pump aft of the first hatch. Peering below, divers can see the offset centerboard — an old shipbuilding technique that was later abandoned. A winch can be seen next to the remnant of the main mast. The mast broke off approximately four to five feet above the deck.

The aft end of the *Oxford* is simply fantastic. Although the cabin is missing, the stern offers plenty for the deep wreck diver. The highlight is the tiller. Approximately ten feet long, it was the precursor to the wheel and should not be overlooked.

The transom is spectacular. The rudder — partially turned towards port — is in phenomenal shape and is still connected to the tiller. The way the silt rests, you can actually see the rudder, transom and partially underneath the ship.

The ten foot long tiller is still attached to the rudder (CRIS KOHL)

Oxford

Prior to identification, the *Oxford* was known as the Tiller Wreck and the Crow's Nest Wreck

13 Persian

Side scan image

Construction

The War of 1812 provided a spark for shipbuilding in the Great Lakes. Carpenters were brought from all over the country to Lake Ontario and Lake Erie in order to build the fleet that would not only see battle against the British, but win the war. These were the unsung heroes.

The lakes were the "great highway" for commerce and trade. This boom created an entirely new industry. Previously, ships were mostly constructed by and for their owners. Now, with shipyards thriving all over the Great Lakes, the industry was open to other entrepreneurs. Cleveland was at the forefront of shipbuilding during this period and the shipyards of Thomas Quayle were leading the way.

Thomas Quayle came to Cleveland from the Isle of Man in 1827 as an apprentice carpenter and shipbuilder. By 1847 he had formed his first partnership and founded Coady, Quayle & Company. Three years later the business was renamed Quayle & Moses. In 1852 he founded the legendary Quayle & Martin Shipbuilders.

Due to his knack for building sturdy, staunch vessels, Quayle's business expanded rapidly. It was not uncommon for his shipyard to have seven

Ship

Official Number 150064
Type Wood propeller
Built 1874
Dimensions 243' x 40' x 19'
  (74 x 12 x 5.8 m)
Tonnage 1630 net tons
Power Steam engine
Builder Quayle & Murphy, Cleveland, OH
Owner Winslow & Winslow, Buffalo, NY
Previous Names N/A
Date of Loss 26 August 1875
Cause Fire
Lives Lost 0
Location GPS 42 33.781 -79 54.696

Dive details

- **Max Depth** 195 feet (59 m)
- **Visibility** 50 feet+ (15 m+)
- **Water Temp** low 40s°F (4–7°C)

Safety
Starboard side of wreck blanketed in fishing nets. Cargo and hull almost completely filled with silt.

★★★ Trimix

hulls under construction at any one time. He even built the infamous *Dean Richmond* (see *Shipwrecks of Lake Erie Volume One* for a complete history of this treasure ship). By 1874 he founded Quayle & Sons. He was now only building massive ships including the *Persian* which, at the time of construction, was the largest freighter ever built on the Great Lakes.

The *Persian* cost an astounding $120,000 (over $2.5m in today's money) to build! Her price tag was not the only large thing about her. She measured over 240 feet in length with a beam of 40 feet. Her holds had a depth of nearly 20 feet. She was built to carry over 1,400 tons of iron ore and 75,000 bushels of wheat.

Globe Iron Works provided the propulsion for the ship. They constructed two high pressure engines along with two boilers, each made of one half inch thick iron and measuring 18 feet by eight and a half feet. It took a lot of power to propel the two screws, which could make upwards of 90 revolutions per minute. Additionally, she was built with four masts — the tallest being 90 feet with a 26 inch diameter.

Everyone knew that the *Persian* was a special ship. Even the press hailed her. On 15 July 1874, the *Buffalo Commercial Advertiser* reported:

"The mammoth steamship *Persian* is about ready to launch at Cleveland, and will probably be placed in the water some day the next week. She is a big-un".

Once she was put into service the build record setting *Persian* immediately began to break shipping records too. She stole headline after headline. This from the same paper, 28 August 1874:

"The mammoth steamer *Persian* has returned from Cleveland on her first round trip with 1,816 tons of gross weight, said to be the largest ever brought from Marquette".

Then on 16 November the same year:

"There arrived in port Saturday afternoon the largest boat with the largest cargo of grain ever brought to this city. This craft is named the *Persian*, and her cargo consisted of 63,159 bushels of wheat; although she has the capacity for several thousand bushels more".

In May 1875 she loaded an astonishing 83,883 bushels of grain in Milwaukee. With the exception of a minor leak in her hull after dropping off this record load she was involved in no incidents. However, that would all change just 13 months after being launched.

Loss and rescues

On 22 August 1875, the *Persian*, under the command of Captain Samuel Flint, departed from Chicago with 50,500 bushels of corn and 17,000 bushels of wheat. There were 19 people manifested on the ship — 16 crew and three passengers — on the trip to Buffalo. The weather was fair throughout the first four days of the trip, with a southeast wind. Conditions were ideal.

On Thursday 26 August, the watchman John Evans and First Mate Thomas Casey were standing on the foredeck going about their business on a calm Lake Erie night. At approximately 2130 hours they issued the cry of "Fire!" — quite possibly the worst thing one can ever hear on a ship. Much to their horror, they witnessed flames bursting through the upper deck above the boilers on the port side. As all hands came on deck, they watched as the blaze rapidly spread.

James Love, the *Persian*'s first engineer, was on duty at the time the fire broke out. As he heard the cry he immediately stopped the main engines. He then turned his attention to the pony engine to engage the pump and hose. However, this engine failed to start the pump. Due to the immense heat Love could not get the valve open. Suddenly, the flames burst into the engine room forcing him onto the deck.

Captain Flint ordered the ship to turn hard-a-port in order to bring the bow into the wind — with the hope of keeping the fire as much aft as possible. Meanwhile, the crew attempted to get the stern pump and hose working, but the heat was too intense. Captain Flint later stated "The fire was so hot and increased so rapidly that we found it impossible". They even tried rigging the forward hose and pump but the hose was too short and would not reach the flames. Of course, the one time they really needed water while on the lake, they could not get it.

After valiantly battling the rapidly spreading fire for nearly an hour, Captain Flint ordered the crew to abandon ship. In his own words:

> "It was now about half past 10 pm. We saw it was impossible to save the barge, the flames increasing with great fury and driving us forward. We then saw it was our only chance to save ourselves. The lifeboats all burned, nothing remained but the hatches to save ourselves upon. Sixteen pieces were thrown overboard, on which we saved the crew of 16 persons and three passengers".

For two hours those 19 who entered the water desperately clung to those 16 hatches, waiting for a miracle. Luckily, on this night, Lake Erie's waters were calm and warm. Shortly after midnight the schooner *Montana* arrived at the scene of the burning freighter. Moments later the tug *Merrick* also came to the stricken crew's rescue. The *Montana* picked up 13 of the 19 floating in the water, the *Merrick* took the rest. Miraculously, all survived. Shortly thereafter, the *Montana* transferred them to the *Merrick*.

The *Merrick* and the 19 saved returned to the *Persian* to ascertain if there was even the slightest chance of saving her. As Captain Flint stated, "We found there was no chance of doing anything". The upper deck was gone — it completely burned from the stern all the way to the foremast. All of the cargo of grain was ablaze. However, they were able to grab the ship's compass and some bedding.

When they were preparing to depart, the propeller *Empire State* under the command of Captain Wright arrived alongside the *Merrick*. Captain Flint conversed with his two counterparts. He begged the *Empire State* to do something, to which Captain Wright responded "I will try". With that, all the men aboard the *Empire State* manned all the pumps and hoses in one last attempt to put out the fire. For nearly three hours they pumped massive quantities of water from three hoses onto the flaming vessel. Unfortunately, they could not make any headway.

In a final desperate maneuver, the *Empire State* attached a line to the burning ship and tried to tow the *Persian* to Long Point in order to beach her — a distance of nearly ten miles. Captain Flint stated they:

> "… could not tow her at all. She would go first one way, then the other, and turn the *Empire State* right around. We worked at her for about two and a half hours when we found that we had made no progress. The fire now burst out in the pilot house and Texas with so much heat that we had to abandon her altogether".

The towline was now engulfed in flames. Captain Flint tried to get the *Empire State* to remain until the *Persian* sank, but Captain Wright would have none of it. He stated he had too many passengers on board and needed to keep his schedule. After the line burned off very close to the *Empire State*, it was decided to let Lake Erie have her. There was no sense in risking another ship and scores of men to attempt to save something which could not be saved.

The Long Point Lighthouse keeper watched the entire event unfold. He said that the *Empire State* was alongside the *Persian* battling the fire. The keeper estimated the *Persian* to be 20 miles distant when she was last seen, with white smoke from the burning corn pouring out of the ship. The *Toronto Globe* reported that "The passengers on the *Empire State* say the burning ship, when her rigging was ablaze, was a beautiful sight".

Steward A M Easy of the *Empire State* described the scene when Flint checked who had made it off the stricken ship:

> "Captain Sam Flint, who was in command of the doomed vessel, called the roll after all had assembled on deck, and not one of the crew was found to be missing, all answering to their names. When this announcement was made, there were many expressions of joy and thankful prayers were offered up to the Almighty for this narrow escape".

The full compliment was taken as far as Point Au Pelee by the *Empire State* where they were transferred to the barge *Anna Smith*. The men were later taken to Cleveland where they landed at 0300 hours on 28 August. They were questioned by marine inspectors and some prepared written statements regarding the incident. They endured a hellish 29 hours on Lake Erie in which they were lucky to come away with their lives. The 30 August 1875 edition of the *Buffalo Morning Express* aptly summarized their experience:

> "The lake was fortunately very smooth or all must have lost their lives".

Captain McKenna, Marine Inspector for the Orient Insurance Company, took the lead in investigating the loss of the *Persian*. He chartered the tug *Sarah E Bryant* to take him to the scene of the disaster at the beginning of September. According to McKenna:

> "All that remained to mark the spot where she sunk is a portion of her rigging and some charred and broken parts".

He ordered the tug to tie a line to some of the rigging sticking above the waterline to see if it was connected to the wreck. It was found to be firmly attached. According to the *Buffalo Courier* the only "relic brought into port by the tug was a floating portion of the companionway".

Despite the fact the *Persian* was not insured, the owners decided not to raise the hull as the cost would have been far more than the charred remains were worth. It was also decided not to raise the boilers or engines even though they were only a year old. Her owners declared her a total loss and abandoned her to the depths of Lake Erie 13 short months after her launch. When she was built she was the largest vessel on the Great Lakes. When she sank she was still the second largest.

Diving the Persian

In June 1996, the Canadian Navy research vessel *Cormorant* embarked on a two day expedition focused on underwater archaeology off Long Point, Ontario. The 248 foot ship manned with a crew of 80 carried a six-man submarine with intricate, sophisticated lights and cameras. Rob Cromwell organized the expedition. He led the submarine to several shipwrecks including a large wooden steamer. They had discovered the elusive *Persian*,

121 years after she foundered. Garry Kozak had side scanned this wreck during his hunt for the *Dean Richmond* in the 1980s (see *Shipwrecks of Lake Erie Volume One*). However, it was not until 1996 that the *Persian* was positively identified. Initially, due to the immensity of this wreck, it was believed to be that of the *Marquette & Bessemer No 2*.

The *Persian* rests at a depth of 195 feet. Needless to say, very specialized training, equipment and experience are needed to dive this wreck. She is one of the deepest wrecks in Lake Erie, and is very seldom dived. There is sometimes a current and parts are draped with fishing nets. Divers beware.

There is extensive fire damage to her. The reports of her burning to her waterline are accurate, as you will see if you dive her, the evidence of an inferno is everywhere. The biggest clue of a fire is that only the lower half of her hull remains. Although she does not have high relief, there is enough to make a worthwhile dive, with the wreck rising approximately six to eight feet above the bottom.

Commencing at the stern divers can view the rudder and one blade of the propeller — the rest being buried. If visibility is good (which it usually is on this dive, averaging at least 50 feet) and you know what you are looking for, then you should be able to see the rudder post, attached to the rudder, sticking out from the hull.

Heading forward from the stern, after a decent swim of 50 feet you will see the remains of the largely intact engine. It cannot be missed as it rises at least ten feet above the wreck. One of the large boilers is present and visible on the port side just forward of the engine.

Continuing forward to the cargo area, there is not much to see as the area is filled with silt and pieces of wood. If you make the long swim to the bow you can see more evidence of the fire — the charred bow stem.

> The *Persian* was once the largest freighter in the Great Lakes

14 Saint James

A diver shines his light on the scroll head

TOM WILSON

Discovery

On 15 June 1982, Garry Kozak was searching for the *Dean Richmond* (see *Shipwrecks of Lake Erie Volume One* for a complete account). He was nearing the end of a ten year hunt for the elusive steamer that would eventually cover a whopping 550 square miles of Lake Erie — astounding, and unheard of at the same time. His search would ultimately lead to the discovery of more than 30 shipwrecks.

On this particular day Garry saw the unmistakable side scan image of a fully intact schooner — bowsprit still in place. It was not the *Richmond*. What made this even more remarkable was that her two masts were still upright. Garry had just discovered his 18th shipwreck and aptly titled it "Wreck 18".

A new or "virgin" shipwreck always attracts an abundance of attention. However, there was even more excitement with this particular one given her pristine condition. She was resting upright in 165 feet of water, with both masts standing 80 feet above her deck as if she was still sailing. A simply magnificent wreck. The exquisite 19th century craftsmanship was evident everywhere: a ram's likeness was carved into the figurehead; deadeyes, belaying pins and mast hoops adorned the wreck. Even though the ship's

Ship

Official Number 22417
Type Two masted schooner
Built 1856
Dimensions 118' x 25' (36 x 7.6 m)
Tonnage 226 gross tons
Power Sail
Builder Merry & Gay, Milan, OH
Owner General Charles M Reed, Erie, PA
Previous Names N/A
Date of Loss Late October 1870
Cause Unknown
Lives Lost 7
Location GPS 42 27.104 -80 07.331

Dive details

- **Max Depth** 165 feet (50 m)
- **Visibility** 40 feet+ (12 m+)
- **Water Temp** upper 30s to low 40s°F (3–6°C)

Safety

Can be dark. The holds are almost completely filled with silt. The cabin should not be penetrated due to the levels of silt. Temperature in the low 40s year round.

 Tech

wooden wheel was partially buried, it was intact. Even the after cabin was still complete, as was her rudder.

Other uncommon features included a capstan. Although a capstan in itself is not unique, this one, mounted amidships, still had the readable engraved manufacturer's nameplate stating "Talcott & Underhill, Oswego, NY".

The wreck also possessed two completely different bilge pumps — suggesting an overhaul some time after she was constructed. One was made of wood and was mounted behind the foremast. The more modern of the two was located behind the mainmast and was a Brokenshire Pump — patented in the 1860s.

Identification

Still the question remained — what ship was this? And maybe more importantly, how did it get there? Until she could be positively identified, she would be known as "Schooner X".

Armed with a plethora of information including the description of the wreck, the capstan plate, accurate measurements obtained by divers and other details, the real detective work began. The key piece of evidence was gathered when divers discovered the tonnage numbers carved into the main beam. This is exactly what maritime historian Art Amos needed to positively identify the wreck as the schooner *Saint James*.

A February 1856 article in the *Cleveland Herald* entitled "VESSEL BUILDING AT MILAN" mentioned a conversation with one of the owners of Merry & Gay Shipyard. In the article he is quoted as saying "Our enterprising neighbors are as active as ever in putting up staunch vessels for the coming season". The *Saint James* was one of nine schooners built in the early part of 1856 at the Milan, Ohio shipyard.

The *Saint James* was built for General Charles M Reed of Erie, Pennsylvania. She measured over 160 feet in length with a beam of 24 feet and a depth of 11 feet. The wooden schooner had two masts, a square stern and a beautifully ornate figurehead.

Sinking and raising

It did not take long for the *Saint James* to be indoctrinated into the rigors of sailing the Great Lakes. In late June 1857, 14 months after her launching, she was traveling down the Saint Mary's River. She was en route to Erie from Marquette, Michigan with a load of 286 tons of Lake Superior iron ore. While she was making her way south, she struck a "ridge of rocks" in the Nebish Rapids and sank in just five minutes!

The 11 July edition of the *Buffalo Daily Republic* reported on the possibilities of raising the wreck. It did not look good. It was written that experts:

> "… think her case is hopeless without the aid of a vessel upon each side of her. She lies very badly upon the crest of a reef of rocks in a current of 10 miles an hour, which is sweeping over part of her deck".

Despite the gloomy forecast for refloating her, as everyone thought she would eventually be torn apart, an attempt was made nonetheless. On 15 July 1857 the *Buffalo Daily Republic* reported that the *Saint James* had been raised with the headline:

"STEAMER ST. JAMES RAISED — QUICK WORK".

The article stated:

"An expedition left Detroit for the purpose of raising her, having two steam pumps for the purpose, one a Worthington belonging to the Buffalo Mutual Insurance Company, in charge of John Berryman, engineer, and the other a Holly Rotary belonging to Bagnall & Dobbins, in charge of George Richardson, engineer. On arriving at the wreck, they turned to, under the supervision of that best of wreckers, John Berryman, and pumped her dry in one hour after starting the pumps".

"They then discharged part of the cargo and cut away the ceiling, when they found a hole 4 by 6 feet in the bottom, and five beams broken".

After securing the hole, the recovered wreck complete with 150 tons of ore was towed by the *Iron City* to Detroit where she was repaired. The cost to raise her was $2,200. She was lucky. However, her luck would not last forever.

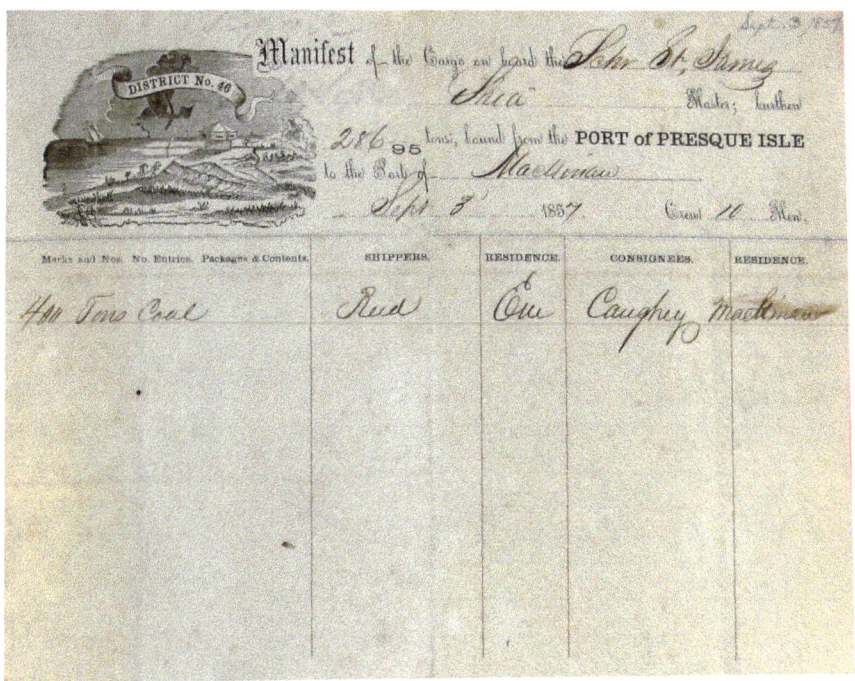

Original manifest from the *Saint James* dated 3 September 1857
(BURTON HISTORICAL COLLECTION, DETROIT PUBLIC LIBRARY)

Final loss

The *Saint James* was involved in several minor incidents after her first sinking. In early March 1860, she was being towed out of Chicago with a load of corn when she collided with the schooner *Thomas Kingsford*. The *Saint James'* jib boom and head gear were carried away — which cost $200 in repairs. The damage to the *Kingsford* was much less expensive at $35.

In April 1861, she went ashore in the Straits of Mackinaw. She was released with minor damage. Finally, in June 1869, the *Saint James* again went ashore, this time in Lake Erie on Point Pelee. It was reported she "got off leaky". She was completely overhauled two months later.

On Friday 28 October 1870, the *Saint James* took on a load of 14,000 bushels of wheat in Toledo, Ohio. Captain John Burrill, with his crew of five men and one cook, set out late that evening bound for Oswego, New York. The trip would take her along the entire length of Lake Erie to Port Colborne, up through the Welland Canal, into Lake Ontario, and then east into Oswego. After she departed Toledo, the *Saint James* was never seen or heard from again.

"What happened?" is the million dollar question. Her sinking is completely shrouded in mystery. There were no survivors. No eyewitnesses. No accounts of attempted rescue or assistance. No wreckage ever washed ashore. No bodies were ever found. No yawl boat was discovered. No lifebelts were found floating in the lake. She literally disappeared without a trace.

Very little was ever written about the *Saint James* loss. In fact, after countless hours researching, I was only able to locate a couple of newspaper articles. One snippet appeared in the 7 November 1870 *Titusville Herald* and was sandwiched between a short piece on private euchre card parties and local train times, under the inauspicious title of "Here and There". It is definitely a paragraph that could be overlooked. It reads:

> "On Friday the 28th ult., the schooner *St James*, Captain Burrill, left Toledo with a cargo of fourteen thousand bushels of wheat consigned to a lower lake port. From the time she sailed out of Toledo harbor, no word has been heard of her crew. She has never arrived at Port Colborne, and the length of time which has elapsed since her departure upon the fatal passage leaves no responsibility but that she foundered with her crew of seven human beings. She was owned in Buffalo, valued at $10,000".

The 9 November edition of the *Toledo Blade* provided slightly more detail:

"Although nothing certain is known concerning her fate, there can now be but little doubt that the worst has happened. The *Saint James*, Captain John Burrill, left Toledo Friday evening, October 28th, with a crew of five men and a cook, laden with 14,000 bushels of wheat, bound for Oswego, since which day nothing has been heard of her, either at Port Colborne or at the port of her departure. Her long absence precludes the slightest hope of her yet being afloat, although she was a staunch craft and overhauled only last August at a cost of $3,500. She measured 226 tons, was valued at $10,000, and insured for $8,000 — equally divided between the Western and Security companies".

The article ended with the following:

"Her commander was one of the most able seamen on the lake, and, sad to relate, leaves a wife and six children to mourn a calamity that can never be repaired".

The question remains — what happened to the *Saint James*? It has been written that she was purposely scuttled because she was an aging ship and was "destined to be scrapped". This is not supported by the facts. The suggestion was that the crew sank the ship and then secretively rowed to shore and stayed in their homes. Apparently, this would not be an uncommon practice for the shipowner Charles Reed. The main problem with this hypothesis is that although Reed was the original owner when the *Saint James* was constructed in 1856, he was no longer the owner in 1870. She had been purchased in May that year by Andrew Abernathy, after being completely overhauled. I am not sure why anyone would want to scrap a newly renovated vessel that they had just purchased.

Since purposeful scuttling is an extremely unlikely explanation, what are the other possibilities for how she foundered? Leak. Storm. Collision. Fire. Or any combination of the four. If one knows how a ship can founder, the next logical thing to consider would be the evidence and to ask, how does the wreck look? Is there any damage? If so, where is it? What does the damage look like? Is anything missing from the wreck? From there the questions progress. What is her position in the water? Is she turtle (upside down)? Is she sitting upright? Once these and other questions get answered it is possible to start narrowing the realm of possibilities down to a plausible explanation.

Let's start going through the list. It does not take an expert to realize there is no fire damage to the *Saint James*. This is easily determined by looking at the wreck. The telltale signs would be everywhere — specifically, heavy charring. There is no sign of charring, therefore the *Saint James* sinking due to a fire is off the table.

I would also strike out the possibility of a collision. There is no evidence anywhere on the wreck to suggest she was involved in an accident. No gaping hole in the hull. None of the gunnels are staved in. Nothing. She is about as pristine as a shipwreck can be.

Now that fire and collision are no longer possibilities, the only other options would be a leak, storm, or a combination of the two. Is it possible to tell the difference? In some circumstances, yes. However, it can be extremely difficult. This may be one mystery that cannot be solved. However, I will present my theory as to what happened.

Some evidence that suggests she slowly sank due to a leak is that the masts are still standing and the aft cabin is completely intact. Typically — not to say always — cabins are ripped off a sinking vessel as the in-rushing water compresses the air inside the ship. The air needs to escape. Many times it escapes by blowing off the superstructure — especially in wooden vessels that sink quickly.

If, in fact, she slowly went below the surface, the next question is "What happened to the crew?" So far as known information is concerned, they never reached shore. No bodies were ever recovered. No bones have been seen on the wreck. What happened to them?

If the seas were calm at the time the *Saint James* foundered, it is difficult to imagine that the crew could not have rowed to shore, or at least remained afloat in their lifeboat. So, I believe that whether at the time of the sinking or sometime after the sinking — while the crew were in a lifeboat — they encountered a severe storm. Typically, in these cases I would research the weather reports for the day she foundered. However, the US Signal Office did not begin publishing the US Daily Weather Maps until 1 January 1871 — two months after the *Saint James* disappeared.

Instead, I turned to the newspapers issued in the weeks after she disappeared. I focused my efforts on those published within one week of 28 October. What I found seems to support my theory. The 1 November 1870 edition of the *Buffalo Express* reported a story about "a severe gale from the East on Sunday night" in which the schooner *William John* was wrecked in Kingston. This "severe gale" occurred on 30 October — less than 48 hours after the *Saint James* departed Toledo.

Another article that caught my attention was published on the same day, in the *Titusville Herald*. It was entitled "Fearful Gale on Lake Erie — Loss of Life and Vessels" and appeared in bold letters. The opening is intriguing and lends more credence to the idea that the *Saint James* or perhaps the crew encountered the terrible storm:

> "We learn from an eyewitness, who arrived from Erie last evening, that one of the most violent gales ever experienced prevailed … resulting in the loss of several vessels and many lives. The gale commenced from S.S.W. to the N.W. about 2 o'clock Sunday afternoon and blew with increasing violence all night".

It is very possible that the *Saint James* and/or her crew foundered in this storm. Given that there is no obvious damage to the wreck, it is feasible that she foundered due to a leak, with the crew abandoning ship prior to the storm commencing. The lifeboat then must have succumbed to the storm. The newspaper reported that the gale started in the afternoon and gradually worsened. This would fit. At this point in history, the fastest crossing of Lake Erie was 16 and a half hours — by steamer. A schooner would be much slower, although it would depend on the winds. It is therefore plausible that the schooner *Saint James* departing Toledo at the western end of the lake very late in the evening or close to midnight on Friday would be off Long Point on Sunday afternoon in the middle of the storm.

Diving the Saint James

Regardless of how she foundered, the wreck of the *Saint James* is spectacular. Sitting upright in 165 feet of water, she is a very photogenic wreck and presents many opportunities for underwater photographers. She is simply magnificent. Starting at the bow — which points north — divers can see the dual fluted anchors. Both the starboard and port anchors are in place just as they would have been when she was under sail. The bowsprit is intact. Beneath the bowsprit is one of the few figureheads in Lake Erie. The beautifully carved scroll is a "must see" and a "must photograph" for those with underwater cameras.

Heading aft, there is a winch and a hatch leading to the chain lockers below. The foremast is in sight at this point — still standing after nearly 150 years on the bottom. The fife rail is intact around the mast. These are not very often seen — they are usually crushed when the mast breaks off.

Continuing aft, two hatches (both covered) and the capstan with the nameplate can be seen. You will reach the second mast, which was standing with a fife rail around it until approximately 2001–2003 when it finally collapsed after a significant lean. It now rests horizontally across the deck. Between the aft mast and the rear cabin is another covered hatch. The ship's wheel looks to be intact although it is half buried in silt. This is another stop for those with cameras. The pristine rails are full of deadeyes and other sailing artifacts. There is an abundance of other nautical paraphernalia strewn about the wreck.

The transom is not be missed and presents another excellent photo opportunity. Even though the wreck is covered in zebra mussels, some of the wood can be seen here. The rudder is still complete and is turned slightly to starboard.

> One of the best-preserved schooners throughout the entire Great Lakes

Side scan sonar image of the *Saint James* (GARRY KOZAK)

15 Sir C T van Straubenzee

Yawl resting against the port stern

Famous name

The *Sir C T van Straubenzee* was launched in 1875. She measured 127 by 26 feet, by 13 feet in depth. The three-masted bark's namesake was Charles Thomas van Straubenzee — an officer who served in the British Army from 1828–1881. Born at Fort Ricasoli on the British-occupied Mediterranean island territory of Malta in 1812, van Straubenzee hailed from an esteemed military family. He distinguished himself at the Battle of Manarajqur while in command of British troops in China and Hong Kong. He was appointed Governor and Commander in Chief of Malta, a post which he held from 1872–1878. He was later promoted to general and remained the head of government until he retired in 1881. Due to his lifetime of service to the Crown, van Straubenzee was awarded the Knight Grand Cross of the Order of Bath. In other terms, he was knighted (by Queen Victoria) and he officially became Sir Charles Thomas van Straubenzee.

Despite his fame I have found only one newspaper story which uses the correct spelling of "Straubenzee". All other articles and books use "Straubenzie" when speaking about the ship. The ship's builder was a big admirer of Straubenzee. I find it hard to believe he would have got the

Ship

Official Number C75632
Type Three masted barquentine
Built 1875
Dimensions 127′8″ x 26′2′ x 12′
(39 x 8 x 3.6 m)
Tonnage 317 gross tons
Power Sail
Builder Louis Shickluna, Saint Catherines, ON, Canada
Owner Pittsburgh & Erie Coal Company, Pittsburgh, PA
Previous Names N/A
Date of Loss 27 September 1909
Cause Collision
Lives Lost 3
Location GPS 42 32.611 -79 55.448

Dive details

- **Max Depth** 200 feet (61 m)
- **Visibility** 50 feet+ (15 m+)
- **Water Temp** upper 30s to low 40s°F (3–6°C)

Safety

Temperatures in the high 30s and low 40s year round. Currents can be present on this wreck. Heavily silted. Beware of fishing nets.

 Trimix

spelling incorrect, especially as he was naming a ship in his honor.

Construction

The *Sir C T van Straubenzee* was built by Louis Shickluna in Saint Catherines, Ontario, Canada, at the Shickluna Shipyard. Commodore Louis Shickluna was a shipbuilder and ship designer responsible for some of the early types of schooners on the Great Lakes. He originated from Malta. A century old newspaper, the *Saint Catherines Evening Journal* described him as having:

> "[The] qualities for which his countrymen are so famous for, excelling in shipbuilding, combined with American goaheaditivenes".

It is not known if Shickluna met van Straubenzee himself; however, he did travel back to his homeland in 1871 and 1872 — the time when the latter was governor of the island nation.

The same paper described Shickluna as a "king amongst shipbuilders", explaining that:

> "The vessels built by him have always proved staunch and seaworthy and may be found in almost every port on the Lakes, the pride of their owners".

Sir C T van Straubenzee

Shickluna Shipyard 1863

An early drawing of the *Straubenzee* from the 1870s

It was precisely this reputation that kept the Shickluna Shipyard constantly busy from the early 1840s well into the post Civil War era. Shickluna employed over 160 men in his yard.

The craftsmanship and steadfastness with which they laid keels was a particular area that they excelled at. The *Toronto Globe* wrote:

> "If [the Shickluna Shipyard] does not bear a worldwide celebrity, it enjoys a North American reputation of being A-1 in shipbuilding".

Quite a statement considering all the many shipyards in the Great Lakes in this period.

The *Sir C T van Straubenzee* was built to ply the waters involved in trade throughout the Great Lakes. Early records show that she carried a variety of cargo including wheat, rye, timber, corn, iron, salt, coal and even ice. She regularly traveled the route from Kingston, Ontario to Chicago, Illinois, carrying those mundane goods which built the towns and ports of the Great Lakes.

Pre-1900 painting (digitally enhanced). Besides one newspaper article, this is the only reference I could locate in which the vessel was referred to by its proper spelling.

Incidents

In August 1893, the *Straubenzee* ran aground off Pigeon Island in Lake Ontario while she was "tacking down and got too close to the shoals". After a tug, a schooner, a derrick and a "gang of men" worked endlessly for two days unloading her, she was finally freed. Even though it was not thought she had suffered any damage or sprung any leaks, she was taken to dry dock where she was caulked. The 2 August 1893 edition of the local paper, the *Daily British Whig,* wrote "She is a very staunch vessel, but has been in hard luck, being in several accidents".

One of the more interesting episodes occurred in August 1878. According to several newspaper reports, on Saturday 24 August 1878, at 2100 hours, a collision occurred approximately 30 miles off Cheboygan, Michigan (between Detour and the Spectacle Reef Light) involving the schooner *Grace Murray* and an unknown Canadian vessel. According to the reports, the *Murray* "was standing to eastward and the other [vessel] to westward, and struck the *Murray* on the quarter, tearing away her cabin".

Unfortunately, during this incident the cook was knocked overboard and the mate was "badly cut about the hand" which required medical assistance at Cheboygan. Sadly, despite a rescue attempt, the cook (a Swede named Nicholas Daly) drowned. The whereabouts and identity of the other vessel were unknown as it did not stop to render aid or assistance. It reads like an 1878 version of a hit-and-run.

Some believed that the other vessel was the *Sir C T van Straubenzee*. The *Grace Murray* filed an official protest in Chicago. As of 31 August 1878, the captain of the *Straubenzee* had yet to be fully interviewed. However, the *Chicago Inter Ocean* reported "It must not be understood that he denies it". Basically, the captain of the *Straubenzee* "took the Fifth". I do think it is telling that he did not deny he was involved in the collision. It stands to reason that if you were not, but were accused of being, you would vehemently deny such a charge. Invoking the Fifth might mean he had something to hide.

The following year, on 13 May, the *Straubenzee* went ashore on Bois Blanc Island, Michigan. The following day she was pulled off by the tug *Bennett*. Two weeks later, she barely averted disaster when the tug *Castle* and schooner *Alleghany* nearly collided. There were five schooners in the *Castle*'s consort — *Halstead*, *Parsons*, *Straubenzee*, *Blake* and *Hamilton*. The *Halstead* and *Parsons* were lucky and avoided collision. However, the other three all collided with the *Alleghany*. The *Hamilton* lost her mainmast, tore her foresail, broke her topsail, broke 18 stanchions on the starboard side, and damaged multiple chain plates. The *Blake* lost her mizzen shroud and had other minor injuries. Whether there was any damage to the *Straubenzee* is unknown.

The next year the ship again avoided certain disaster due to a remarkable feat of bravery by the captain of another vessel. On Saturday 23 October 1880, three schooners were caught in a storm on Lake Ontario. The *Blake* and *Gulnare* sailed past the lighthouse at Port Dalhousie, Ontario, over the pier and went ashore. The *Straubenzee* was the next schooner in line and was headed for the rocks and beach. Alex Milligan, captain of the tug *Richardson*, was watching the *Straubenzee* heading towards the pier. The only way to save her was to get a line to her and attempt to pull her into port before she plowed into it. With no other means of doing so, Captain Milligan tied a line around his waist, jumped off the pier into a heavy sea and swam out to the *Straubenzee*. He untied the line from his waist, handed it to the crew and swam back to his tug! Remarkably, the *Richardson* was able to tow the *Straubenzee* into port and save her from a certain collision with the pier. An astounding display of heroism.

For the next three years the *Straubenzee* sailed without incident. Then, in November 1883, she went ashore at Point Prophey, Lake Superior. Unfortunately, there was no Captain Milligan to save her this time. Fortunately, she only suffered a broken rudder.

In December 1889, one of the more bizarre incidents involving the *Straubenzee* occurred in the port of Oswego, New York, on Lake Ontario's southern border. According to the "Marine Intelligence Report" in the 2 December 1889 edition of the *Daily British Whig*:

> "A dozen sailors under the influence of liquor boarded the schooner *Sir C.T. van Straubenzee*, at Oswego, and demanded that the second mate, F. Talmage, [of] Picton, desert the vessel".

The drunken sailors said he was "a scab and unworthy of living". The sauced-up crew crowded into the main cabin and demanded that the captain turn him over. After the captain telephoned the police, the drunkards disbanded. However, they regrouped at the grain elevators, demanding the *Straubenzee* surrender the second mate. The outcome is unknown. Still it is a fascinating lost account of Great Lakes lore.

On 6 November 1897, the steamer *Idaho* encountered one of the worst storms to hit Lake Erie that year. Only two of the 21 on board survived. A witness to her sinking, Frank D Root, captain of the steamer *Mariposa*, stated "I never saw old Lake Erie in a worse mood than she was from early Friday morning until I got in tonight". In an attempt to deflect blame for the incident, Western Transit Corporation manager Mr Douglass stated:

> "In all probability nobody will be blamed for the disaster. The men who were in charge of the boat at the time she foundered are all lying at the bottom of Lake Erie".

The *Straubenzee* picked up one of the *Idaho*'s lifeboats. It was empty and "looked as if it had been handled pretty roughly". Still the crew recovered and delivered it to port — a sombre reminder of how dangerous Lake Erie could be.

Collision and sinking

For the next twelve years the *Straubenzee* went about her business of hauling cargo throughout the Great Lakes, largely without making headlines. Then on 27 September 1909, they read:

> "SCHOONER SUNK, THREE PEOPLE OF THE CREW DROWNED"

> "TWO SEAMEN WERE PULLED FROM HER WRECKAGE".

The *Straubenzee* had collided with the palatial steamship *City of Erie*.
At one time the largest side-wheel steamer in the lakes, the *City of Erie* was also one of the staunchest built vessels. Referred to as "practicably unsinkable" due to her hefty construction, she was legendary not only for her strength, but also for her speed. In 1901, she defeated the steamer *Tashmoo* in a memorable and highly publicized race between the ports of Cleveland and Erie — she won by a mere 45 seconds!
The *City of Erie* operated nightly between Cleveland and Buffalo. She departed on her regular trip at approximately 2100 hours on 26 September 1909. She was scheduled to arrive the following morning at 0730 hours. It was raining and the wind was from the northeast. Once outside the confines of the Cleveland breakwater, Captain McAlpine turned over control of the steamship to her pilot, Edward S Pickel. Once they reached Fairport Harbor, Ohio, Pickel plotted a course further north than usual due to the seas which

were coming from the north, so as not to be pushed too far south away from the shipping lane.

Meanwhile, the *Straubenzee* departed Port Colborne, Ontario, bound for Cleveland with five crew members aboard. She entered Lake Erie at approximately 2330 hours. The wind was from the north.

At 0237 hours in the early morning of 27 September, the *City of Erie* passed Long Point, Ontario. Shortly thereafter the primary lookout relieved the regular wheelsman at the helm so he could go below decks. An extra watchman took the place of the regular lookout. Pilot Pickel was still in command of the wheelhouse.

A brief time after the changing of the lookouts, Pickel peered through his looking glasses and saw a red (port) light about a point-and-a-half off the *City of Erie*'s starboard bow. He did not, however, see a green (starboard) light. He thought there was a sailing vessel on the *City of Erie*'s course.

Pickel immediately gave an order to the wheelsman to port the vessel in order to go under the sailing vessel's stern. When the maneuver was executed, the bearing of the red light did not change fast enough to avoid a collision. As a result, Pickel ordered the *City of Erie* hard aport in order to make the steamer swing faster. Suddenly, Pickel "perceived a dark shape take form out of the night". The red light of the sailing vessel was now dead ahead and then the light completely disappeared. Pickel signaled the engine room to stop the engines. He then ordered the engines full astern. At this time, the vessels were only a half mile apart. It was too late to avoid a collision — the *City of Erie*'s forward momentum would prevent her stopping in time. A voice was heard on the sailing ship screaming "Hard up! Hard up!" Seconds later, at approximately 0300 hours, the *City of Erie* plowed into the *Sir C T van Straubenzee* on her starboard side near the main rigging.

Once the *City of Erie* pulled away, the *Straubenzee* began to fill with water and sank almost immediately. The *City of Erie* lowered her lifeboats in an attempt to save the *Straubenzee*'s crew. The captain, mate James McCallum, and the cook Madeline Connolly were never seen again.

Straubenzee crewman William Thomas Garner woke up when he heard the captain screaming "Hard up! Hard up!" and ran out onto the deck. He felt the vessel starting to settle and jumped onto the fore-rigging and started climbing. As the ship slipped beneath the water, Garner swam up to the surface and was saved by a *City of Erie* lifeboat. The only other crewman rescued was Thomas Hollis, who followed Garner to the deck. Hollis jumped onto the main rigging. After screaming "I can't hold on much longer" while he was struggling to keep his head above the water, he yelled, "Hurry up fellows. I'm slipping!" Before he went below the surface forever, those in the lifeboat plucked him out of the water.

The *City of Erie* stayed in the vicinity of the wreck until dawn in order to search for the missing crew members. After all hope was lost, she continued her journey to Buffalo, where she arrived at port at approximately 0910 hours and reported the accident.

Postcard of the *City of Erie* (1911). Her collision with the *Straubenzee* resulted in the deaths of three sailors.

Aftermath

Afterwards, the main point of contention was whether or not the *Straubenzee* had had her green light burning. The matter was swiftly referred to the Supervising Inspector General for the United States Steamboat Inspection Service — Captain James Stone. Multiple members of the crew of the *City of Erie*, in addition to her captain and pilot, were interviewed extensively. They all emphatically stated that the green light was not visible at any time prior to the collision. General Freight Agent H R Rogers of the Cleveland and Buffalo Transit Company (owner of the *City of Erie*) was so certain they would be exonerated that he issued the following statement:

"I am confident the collision was due to the negligence of the schooner in not properly displaying her lights. I know the night was clear and that the collision could have been avoided easily if the other boat had been seen in time. But the *City of Erie* has a speed of a mile in four minutes, and when the obstacle was visible at a distance of only half a mile, it was practically impossible to turn out of the course in time".

The only member of the crew of the *Straubenzee* to testify was William Thomas Garner. He stated that while he was climbing the fore-rigging he climbed over the green light and noticed it was burning. Several men from the Port Colborne area were in agreement with Garner. They steadfastly held that both the red and green lights were burning on the *Straubenzee* when they towed her out of the breakwater and into Lake Erie at approximately 2330 hours on 26 September.

Despite these statements to the contrary, Captain James Stone ruled that the *Straubenzee*'s starboard (green) light was not ignited at the time of the accident. Therefore, he found the *City of Erie* was not at fault for the sinking.

As is typical when two vessels collide on the Great Lakes, a libel suit was filed, by the Pittsburgh & Erie Coal Company against the Cleveland and Buffalo Transit Company in the United States District Court for the Northern District of Ohio. Judge Day ruled that the starboard light was not burning on the *Straubenzee*. He opined:

"The fact that the red light was shut out would seem to indicate that the schooner was not holding her course, but that she starboarded and swung in the same direction as the *City of Erie* was swinging".

And with the stroke of a pen, the libel suit was tossed. The court placed the blame on the men at the bottom of Lake Erie.

Diving the Sir C T van Straubenzee

The *Sir C T van Straubenzee* was lost to Great Lakes history for 73 years before Garry Kozak discovered the wreck on 9 July 1982 (see page 122).

At a depth of 200 feet, specialized equipment, training and most importantly experience are required to see Lake Erie's deepest shipwreck — she rests well beyond the sport diving limit. Even though the ship is broken into two distinct pieces, the wreck is phenomenal.

Side scan sonar image (GARRY KOZAK)

Starting at the stern, the impressive large metal wheel can be seen, with the steering gear behind it. Forward of the wheel is a pile of rubble where the cabin was once located. A portion of the mizzen mast is visible here. Along the transom are two square hatchways — or possibly doorways — one on the port side and the other on the starboard side. I do not know where they lead, however, they are a rare feature seldom seen in the Great Lakes.

Along the port side of the stern, adjacent to where the cabin was once located, a yawl boat, partially buried in the bottom of the lake, leans against the gunwale. The stern of the yawl is exposed above the mud line and is a "must see" if you dive this wreck. A yawl is quite a rare sight on any wreck as they were usually launched prior to sinking, or floated away as the ship foundered. It is unusual for one to come to rest adjacent to a wreck. A block and tackle can be seen hanging from the yawl's stern. The remnants of its rudder, as well as the delicate tiller, are visible. This is an essential stop for underwater photographers.

Moving forward of the cabin the remains of the aft mast can be seen, broken off approximately four to five feet above the deck. Several deadeyes are visible on the starboard gunwale.

Continuing forward of the aft mast you will encounter the area where the ship is broken in two. The gap between the aft section and the bow section is approximately 25 feet.

The bow section is superb and offers a plethora of nautical items. Here the middle mast rests diagonally from starboard to port. The forward mast stands as if it is still sailing — quite remarkable after nearly 110 years underwater. This is the mast crew member William Thomas Garner climbed to escape the sinking vessel. If you get great visibility, a photo of the still-standing mast against a clear backdrop of water blending from dark to light as the mast reaches towards the surface is essential for your underwater photography portfolio. The foremast fife rail is even intact.

At the bow there are several more artifacts to explore including a small windlass, the starboard anchor and a bowsprit which was once intact but has since fallen. There are some fishing nets and some wire rigging on the deck. Be vigilant so you do not become entangled.

The ship's bell was generously placed on the foredeck for everyone to see. Whoever put it there even removed all the zebra mussels. Seeing a bell — the icon of the ship and the most sought after artifact on a shipwreck — is remarkable. In no other waters on this planet would a bell rest so comfortably on the deck without fear of being removed — a testament to Great Lakes wreck divers. Just in case you are tempted, remember that it is illegal to remove any items from shipwrecks in Lake Erie. Should anyone do so, they risk prosecution and at least a fine and confiscation of dive gear.

The deepest wreck in the lake

16 Smith

Intact helm seen through the wheelhouse window frames

Construction and launch

Eighteen-eighty-one was an interesting year in the post-Civil War era of the United States. The 20th President, James A Garfield was assassinated, the infamous outlaw Billy The Kid ravaged the West before being killed, and the legendary shootout at the O K Corral involving Doc Holiday and Wyatt Earp took place in Tombstone, Arizona. The year also saw the founding of the American Red Cross and Barnum & Bailey's Circus debuted the "Greatest Show On Earth". Meanwhile, the Union Dry Dock Company was busy building a pleasure craft.

Under the supervision of master carpenter John Humble, the wooden three deck passenger cruiser *Albert J Wright* was constructed for Labon B Fortier. Powered by a small boiler and steam engine salvaged from another vessel, the *Wright* measured 115 feet in length with a beam of just over 21 feet. She was built with the keel rising gently to the bow in order to provide extra strength. She was a sturdy vessel.

The day of her launching was quite the occasion. After being advertised in the local papers, the docks were crowded with spectators, press and local celebrities. The rich and the poor gathered together to see the

Ship

Official Number C138371
Type Tug
Built 1881
Dimensions 120' x 22' x 10'
 (37 x 7 x 3 m)
Tonnage 218 gross tons
Power Compound engine
Builder Union Dry Dock Company,
 Buffalo, NY
Owner Sinmac Company, Montreal,
 QC, Canada
Previous Names *Albert J Wright*
Date of Loss 25 October 1930
Cause Storm
Lives Lost 0
Location GPS 42 28.486 -79 59.061

Dive details

- **Max Depth** 165 feet (50 m)
- **Visibility** 40 feet+ (12 m+)
- **Water Temp** upper 30s to low 40s°F
 (3–7°C)

Safety

Temperatures in the low 40s year round. Wheelhouse can be penetrated but is heavily silted.

★★ Tech

newest vessel built by the proud city of Buffalo, New York. The *Buffalo Morning Express* reported:

"The launch was set for between 4:00 and 5:00, and at 4:35, everything being in readiness, Mr. John Humble, master builder for Mills & Co., the builders, gave the word, the last block was knocked out, and the *Wright* glided gracefully down the ways, and not withstanding that she had a drop of about 3 ½ feet from the ways to the water in the dock. She made it easily, both ends touching the water at the same moment, and she settled majestically while a big wave swept over the sides of the dock and some distance over the yard, causing those in the immediate vicinity to step back somewhat hastily in order to escape wetting".

Although launched on 14 May 1881, she would not make her inaugural voyage until a month later. The 13 June 1881 edition of the *Cleveland Herald* printed the following:

"The new steamer *Albert J Wright* made her trial trip Saturday afternoon. She is an unusually staunch and comfortable boat and is commanded by Captain James Doyle. She made her first excursion trip down [the] Niagara River today with a large party on board".

Captain Doyle had an impeccable reputation on the lakes and all on board enjoyed his company. According to one local newspaper, the captain was "extremely popular, especially with the ladies".
The *Buffalo Morning Express* reported:

"As Captain Doyle her commander, is a deservedly popular master, and the boat herself is fitted out with every regard to the comfort and safety of passengers, we have no doubt she will soon become a source of profit to her owner".

The *Smith* in 1929

Working life

Fortunately, Fortier did not have to wait long to start reaping the benefits of ownership. In fact, she was so popular her early years were dedicated to the burgeoning sightseeing and cruising business. At a modest 50 cent fare for an adult and half that per child, her decks were regularly filled with upwards of 700 paying customers for her daily excursions out of Buffalo along the Niagara River. She also had a "handsomely fitted cabin for the accommodation of lady passengers". She later took fare payers on scenic tours out of Chicago. In 1886, she hauled a circus company around the Great Lakes for the summer.

Conversions

Sometime between 1886 and 1888, after being sold to C D Thompson, she was converted from a passenger cruiser to a wrecking tug and was substantially rebuilt. In addition to being outfitted to conduct towing, her mast and third deck were removed. She would no longer ferry people on pleasure cruises, but would instead tow barges and remove navigation hazards throughout the Great Lakes.

At the end of December 1892, the *Wright* headed to Grand Haven, Michigan for the winter in order to have her machinery repaired. She was in her berth on 5 January 1893 when a fire broke out at approximately 1700 hours. The *Detroit Tribune* reported: "A heavy northwest gale was blowing, and before the fire department could get to her, she burned nearly to the water's edge". To make matters worse, she was only partially insured and was considered a total loss with damage exceeding a value of $20,000.

Despite this her owners tried to recoup some money. The following month her engines were salvaged and placed into the *I N Foster* — a converted steamer. She was later sold to Leatham & Smith Towing & Wrecking Company who had plans to return her to towing service. The *Wright* was rebuilt and furnished with old engines from the *City of Charlevoix* and boilers salvaged from the wrecked steamer *Calumet*.

After sitting idle the entire 1899 season, the wrecking tug was converted into "a tug to be used in towing stone and other scows and lighters". She was once again completely rebuilt — receiving new engines and boilers. Six years later, additional significant renovations were done — two masts were added, her second deck was removed and she was renamed *Smith*.

After spending the next 30 years throughout the Great Lakes, the *Smith* was sold to Sincennes McNaughton Company (Sinmac) in Montreal, Canada. In need of another overhaul, she was taken to Sarnia, Ontario in tow of the *Manistique*. Both vessels departed Port Colborne, Ontario on Friday 24 October 1930. In the early morning hours of 25 October, while the clock was nearing 0200 hours, the *Smith* began to sink in stormy seas. The four crewmen aboard were quickly rescued by the *Manistique*. No lives were lost.

The wheelhouse, c.1910

Diving the Smith

Today, the tug *Smith* rests upright at a depth of 165 feet, with a slight list to port. There is typically exceptional visibility on this wreck — sometimes approaching 80 feet or more. When it is that good, the wreck can be seen from as shallow as 100 feet.

At the bow — which points east — a windlass can be seen with the chain still attached to an anchor on the port side. The anchor is cinched tight into the hawsepipe. Just aft of the windlass is a small companionway with some stairs leading below decks. At one time there was space to peer below, although she is now filled with silt and mud.

Heading aft from the bow, a diver will encounter the tug's well-preserved wheelhouse. Inside, although she is partially filled with silt, the intact helm can be seen protruding above the mud. This is a rare sight and a "must stop" for any underwater photographer. Entry can be made into the wheelhouse through an open port door, however, the door is mostly blocked by high silt.

Behind the wheelhouse, resting off the port side, is the fallen smokestack. An observant diver will notice the steam whistle still attached to the smokestack. Continuing aft, most of the superstructure of the engine house has collapsed. The tug's compound engine is visible. Off the rear port side of the engine room is a lone lifeboat davit.

The stern is rounded and remains above the mud. Divers sometimes clear the zebra mussels to expose the ship's name and port emblazoned across the transom: "SMITH — MONTREAL".

An 1882 advertisement for the Union Dry Dock Company from *Beeson's Marine Directory for the Northwestern Lakes*

The highlight of this dive is the intact helm inside the wheelhouse

17 Stern Castle

One of the deepest wrecks in Lake Erie, this unidentified two masted schooner rests at 185 feet. Unlike some of the other unknown wrecks in the Eastern Basin, this one is visited by dive charters. She is dubbed "Stern Castle" due to her most prominent feature — a high-relief stern.

Sitting upright in deep water the bow and stern both rise out of the bottom, while her midsection remains almost completely buried. For underwater photographers this wreck offers several attractive features — both her masts are still standing, the ship's wheel is still intact and in place, there is a windlass, and some of her running gear is visible. Photography is enhanced by outstanding visibility, which often exceeds 50 feet.

Ship

Official Number Unknown
Type Two masted schooner
Built Unknown
Dimensions Unknown
Tonnage Unknown
Power Sail
Builder Unknown
Owner Unknown
Previous Names Unknown
Date of Loss Unknown
Cause Unknown
Lives Lost Unknown
Location GPS 42 30.294 -80 02.379

Dive details

- **Max Depth** 185 feet (56 m)
- **Visibility** 50 feet+ (15 m+)
- **Water Temp** upper 30s to low 40s°F (3–7°C)

Safety

Wreck is heavily silted covering most of the midsection, which can be disorienting if not familiar with the site.

★★★ Trimix

Divers must visit the ship's wheel and her two standing masts

18 Swallow

The Swallow taking on a load of lumber in the late 1870s

Construction

There are thousands of shipwrecks in Lake Erie — vessels which for one reason or another succumbed to the treacherous waters and sank to the bottom. There are only a handful which had the misfortune of sinking twice. There are even fewer ships which have foundered not once, not twice, but three times. The *Swallow* is one of those vessels. Her story began in 1873.

While the Indian War was being waged by General George Armstrong Custer in the west, and amid the Tammany Hall scandals in the east, the A A Turner Shipyard in the small town of Trenton, Michigan was busy building a ship for D Whitney Jr of Detroit. Without a name on the books while the wooden lumber hauler (also known as a lumber hooker) was being constructed, a small bird built a nest in a corner of the vessel's hold. It was decided to name her *Swallow*. The nest remained in place until she was launched.

The *Swallow* had two masts and one smokestack. She had forward and aft superstructure with a sunken deck between which would allow the lumber to be stacked high.

Located just down river from Detroit, A A Turner had his shipyard in

Ship

Official Number 115184
Type Wood propeller
Built 1873
Dimensions 133'8" x 25'8" x 10'8" (40 x 8 x 3.3 m)
Tonnage 256.67 gross tons
Power Steam engine
Builder A A Turner, Trenton, MI
Owner Siebert, Quinlan, James & Lennae
Previous Names N/A
Date of Loss 19 October 1901
Cause Storm
Lives Lost 0
Location GPS 42 34.892 -79 56.455

Dive details

- **Max Depth** 190 feet (58 m)
- **Visibility** 50 feet+ (15 m+)
- **Water Temp** upper 30s to low 40s°F (3–7°C)

Safety

Temperatures in the low 40s year round. Penetration of the engine and machinery spaces under the deck is not advised as the deck above is collapsing. Fishing net attached to the top of the mast.

★★★ **Trimix**

Trenton since 1866. Considered a master shipbuilder, he constructed 36 vessels between 1866 and 1873 including 18 steamers and five lumber barges (one of which was the *Swallow*). The vessels ranged in price from $25,000 to $60,000. At the shipyard's peak, he employed 350 men at a total cost of $3,000 in weekly wages. He once had five vessels under construction simultaneously. He lost his fortune during the Panic of 1873 financial crisis. Sadly, his vessels shared his bad luck, with four being stranded, five burning, and eight wrecking. Just under half of the vessels he built would go on to be involved in tragedy.

When she was launched the *Detroit Free Press* described the *Swallow* as "elegantly fitted out" and said that she "presents a fine appearance and is considered one of the best propellers in the canal trade on the Lakes". Quite a statement for a non-glamorous vessel carrying 300,000 feet of lumber.

Problems begin

Less than two years after her launch, the *Swallow* was involved in her first accident. On the first Tuesday in May 1875, she was at her dock in Toledo, Ohio. Her crew were closing the hatches and preparing her for departure. She had just taken on a load of lumber. Meanwhile, the enormous steamer *Fred Kelley* struck a sandbar near the Lake Shore Elevators across from the *Swallow*'s berth. In an attempt to free that steamer, the engineer of the *Kelley* "pulled the throttle wide open". Fortunately, she gained enough power that she was able to pass over the sandbar. Unfortunately, she picked up so much speed that she couldn't stop before plowing into the *Swallow*. She struck with such ferocity that a hole was staved amidships just below the *Swallow*'s waterline. She was repaired and put back in service.

For the next 30 months she would ply the waters of the Great Lakes, moving lumber from port to port without incident. Then in 1877 a massive winter storm swept across the lakes. A report from the *Cleveland Herald* published 5 November reported that:

> "The heavy wind of the last three days continued in a hurricane last night. Dispatches from Lake Erie, Lake Michigan, and Lake Ontario show that damage to shipping have been numerous and severe".

After the storm was over, eight vessels were beached, including the *Swallow* near Port Stanley, Ontario, and one vessel sank with the loss of one life. Other storms have taken a much bigger toll but this one was a reminder of how dangerous Lake Erie can be.

The next time the *Swallow* made the news was when she crushed the tug *Butler* against the piers of the Rush Street Bridge in Chicago. She was repaired in dry dock and placed back in service.

Three years later, in 1886, the *Swallow* would sink for the first time. The June 3 edition of the *Buffalo Commercial Advertiser* succinctly reported:

> "The steambarge *Swallow* left Muskegon for Chicago yesterday laden with lumber. When in midlake she encountered a dense fog and narrowly escaped collision with a sail vessel by the captain pulling his wheel hard-a-starboard. The sudden move threw the vessel on her beam ends. At the same time a fierce squall struck her broadside and the seas rushed into her hold, putting out the fires. The vessel filled to her decks and would have gone to the bottom had she been loaded with coal or ore. The men, however, were saved".

The *Swallow* drifted in the high seas for nearly four hours, being battered by waves and slowly sinking, before she was spotted by a passing vessel. She was towed into Chicago where she sank in the entrance to the harbor.

She was lucky to have been seen by a ship which could tow her.

This accident could have been prevented if the *Swallow* had not been overloaded with lumber. Unfortunately, in 1886 there were no regulations governing the loading of vessels. The captain determined when the ship was at capacity. He had the final say. If the captain wanted a few extra tons of cargo placed aboard, it was placed aboard. No questions asked.

The fallout from the sinking of the *Swallow* was summarized in the 10 June edition of the *Saginaw Courier*:

> "The wreck of the steambarge *Swallow* at Chicago, which was occasioned by overloading, has drawn attention at Chicago to the general subject of loading a vessel beyond her capacity and a movement to have some regulations made to prevent the practice is on foot [sic]".

The accident may have provoked a conversation, but regulations were not enacted until the 20th century.

In 1891, a year after being sold to Innes, Duncan & Cowan of Chicago for $12,000, the *Swallow* grounded for a second time. She filled with water after running ashore at Michael's Bay, Manitoulin Island, Ontario, Canada, in Lake Huron. She was pumped free of water and finished her journey to Buffalo with no further interruptions.

On Thursday 4 October 1894, the *Swallow* and her consort *Dacotah* were caught in a "furious gale in Lake Erie". The towline between the two vessels parted in the storm. The ships were separated off Middle Sister Island in the western basin. According to the *Buffalo Enquirer* the *Swallow* arrived in Buffalo "in a sinking condition". The *Dacotah* was feared to have been lost along with the seven sailors aboard her, but miraculously she survived. Once again, the *Swallow* had tempted fate and been fortunate. Her luck would not last much longer.

The *Swallow* was involved in another incident in the fall of 1900. This time the headlines read: "SUNK BY AN UNKNOWN!" At approximately 0420 hours on 5 October, she and her consort *Manitou* were traveling down the Saint Clair River from Muskegon loaded with cedar shingles and posts bound for Detroit. Three boats were on an opposite course heading upriver. The *Swallow*'s captain, W P Quinlan, stated that all three were abreast on his port side. He said that his first mate gave one blast of the whistle. However, the call was answered by two whistles from the vessel nearest the *Swallow* (a single toot was the expected response if they were in agreement about the passing situation). Each vessel repeated its signal. The next sound heard was not a whistle, but that of the unknown steel-hulled vessel ripping a ten foot hole in the *Swallow*'s starboard bow. A subsequent report by the US Steamboat Inspection Service stated:

> "The unknown boat kept on her course and didn't stop an instant to see what damage had been done".

The *Swallow*'s crew decided to beach the ship and headed for shore. They never made it, but they were able to get her to the channel's bank where she sank in shallow water. This was a testament to the crew who managed to fight against all the odds with massive amounts of water filling the ship and were still able to steer her close to shore. Remarkably no one was even injured. The results could have been devastating.

Once on shore Captain Quinlan protested and maintained he had the right of way because he was downbound (heading towards the Atlantic Ocean) and that:

> "The other, by his course, has plainly violated the government statutes governing the passage of boats in narrow channels incorporated in what is known as the White Law".

He placed the matter in the hands of his attorneys, Gray & Gray. The Steamboat Inspection Service later determined that the unknown boat which collided with the *Swallow* was the steamer *Sir William Siemens*.

Declared a "total wreck" and "too far gone to warrant raising" it was decided only the cargo of the *Swallow* would be salvaged. The *Swallow*'s own consort *Dacotah* lightered (transferred) the cargo of shingles and posts. The soaked lumber would only fetch $2,000 in Detroit, well below the total loss endured. Meanwhile, the crew "managed to fish their effects out through a hole in the cabin roof, which [was] just above the water".

It was ultimately decided to raise the *Swallow* from the Saint Clair River. Captain John Quinn and his wrecking crew were hired to re-float the steam barge. They began their painstaking work four days after she had sunk. Within seven days, Captain Quinn and his men had successfully brought her back to the surface, floated her down river to Detroit, and pumped her free of water.

Final sinking

The unlucky *Swallow* was next rebuilt and sold to William E Lennae of Detroit, in May 1901. Her bad luck would transfer to her new owners — Lennae would only own her for five months before she sank for the third and final time.

On Sunday 13 October 1901, the *Swallow* took on a load of lumber from the mills at Emerson in northern Michigan, Lake Superior. The crew was comprised of eleven men. Weather was favorable for her journey to Buffalo. The *Swallow* picked up her consort *Manitou* at Sault Ste Marie and then continued on her way.

The two vessels had an easy passage until they ran into a gale on Friday 17 October off Long Point, Ontario. The *Duluth News Tribune* reported:

"The *Swallow* and her tow were off Long Point last Friday night when the gale was at its height. The wind was square abeam and the steamer shipped great quantities of water from every wave that struck her".

Interestingly, the *Swallow* did not cut the *Manitou* loose, as is typical in storms. With her still in tow she battled the waves for hours, putting great stress on both vessels. At approximately 0200 on Saturday 18 October there were over three feet of water in the *Swallow*'s engine room. Her fires were put out. She was sinking. At that point Captain Quinlan decided to abandon ship. After the crew gave the distress signal, the captain finally gave the order to sever the towline between the two vessels. The crew lowered the yawl boat and boarded it. Their only chance at survival was to make it to the *Manitou*. They watched the oil lights still burning brightly on the *Swallow* as they drifted into the darkness.

As they pushed away from the *Swallow*, the deck load of lumber shifted and "came tumbling down about the yawl, but not enough of the timber struck the craft to swamp it". To say the crew was lucky that the lifeboat was not swamped and nobody was injured is a bit of an understatement.

After "a fierce battle with the waves" the men were able to reach the *Manitou*. But their trouble was only just beginning. The *Manitou* had lost both of her masts. She had lost her entire deck load of ties. She was leaking badly. She was now at the mercy of the Inland Seas. She drifted helplessly for the next 39 hours, finally being met by the steamer *Walter Scranton* at 2100 on Sunday 20 October. She towed the exhausted men into Erie.

Diving the Swallow

The *Swallow* drifted in the rough seas and much of her superstructure would burn due to unattended oil lamps when the crew abandoned ship. She sank beneath the surface approximately five miles east of Long Point in 190 feet of water. Divers can expect cold bottom temperatures along with excellent visibility on the wreck. She sits upright with high relief.

Starting forward, much of her foredeck has collapsed. Her pilothouse is missing — most likely ripped off when she sank. Her foremast still stands above the deck and the port anchor can be seen.

Heading aft divers can reach the cargo deck with her three hatches. Some penetration can be done here. The wreck is heavily silted.

Continuing towards the stern, the aft deck has collapsed into a pile on the deck. The single smokestack has fallen on the stern. The curved fantail can be viewed as she sits high above the bottom. The port railing has disappeared, but the starboard railing is intact. A set of stairs can be seen leading below the aft deck near the fantail. Part of the rudder post sticks out of the raised stern deck. A capstan is visible nearby and a portion of the engine and boiler can also be seen.

> The *Swallow* had a disastrous history which included four major incidents and three sinkings

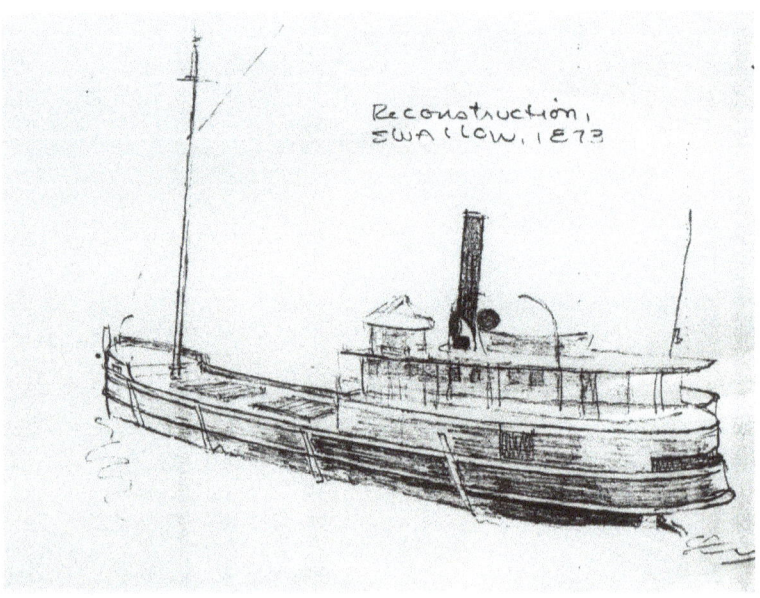

An 1873 drawing of the *Swallow*

19 T-8

The unidentified wooden schooner known as "T-8" is a very rarely dived wreck. Even though it sounds like a droid from *Star Wars*, the "T-8" is actually named after Target 8 — the eighth mark of interest identified by those who rediscovered her in 1995 with a side scan sonar. She was originally found in the 1960s and was named the "22 and a ½ Fathom Wreck" at that time.

Beware of entanglement as the wreck is heavily draped with trawler nets and fishing net floats. The hazard is very real. In 1964, two salvage divers, Tony Lama and Ross Schrum were working to free a net from this wreck. Although visibility is known to be poor here, it was much worse prior to the introduction of zebra mussels to the lake. On that particular day, Lama and Schrum were working in near zero visibility when Lama became trapped in the net. To make matters worse, he lost his mask. Schrum worked hard to untangle Lama. They both made a free ascent to the surface, lucky to come away with their lives.

Both the bow and stern are badly broken. The bow is heavily silted over. The stern, heavily damaged, is hardly recognizable in places. The hull condition suggests she foundered in a collision.

Ship

Official Number Unknown
Type Two masted schooner
Built Unknown
Dimensions Unknown
Tonnage Unknown
Power Sail
Builder Unknown
Owner Unknown
Previous Names Unknown
Date of Loss Unknown
Cause Unknown
Lives Lost Unknown
Location GPS 42 35.226 -80 01.335

Dive details

- **Max Depth** 145 feet (44 m)
- **Visibility** 20–40 feet (6–12 m)
- **Water Temp** low to mid 40s°F (5–7°C)

Safety

Abundance of fishing nets all over the wreck. Net floats raise the nets and increase the entrapment hazard.

★★ Tech

The railing is almost completely intact and provides a point of reference for those circumnavigating the wreck. Divers can also use the intact centerboard.

> The conditions on this heavily damaged sailing vessel make this a challenging dive

Suggested reading

Books

Beeson, Harvey C (1882–1921) *Beeson's Marine Directory of the Northwestern Lakes*. Chicago, IL: Harvey C Beeson.
    (1891) *Beeson's Sailor's Handbook and Inland Marine Guide*. Detroit, MI: Harvey C Beeson.
Bourrie, Mark (1995) *Ninety Fathoms Down: Canadian Stories of the Great Lakes*. Toronto, ON, CA: Dundurn.
Bowen, Dana Thomas (1946) *Memories of the Lakes*. Cleveland, OH: Freshwater Press.
    (1952) *Shipwrecks of the Lakes*. Cleveland, OH: Freshwater Press.
Boyer, Dwight (1968) *Ghost Ships of the Great Lakes*. Cleveland, OH: Freshwater Press.
    (1966) *Great Stories of the Great Lakes*. Cleveland, OH: Freshwater Press.
    (1971) *True Tales of the Great Lakes*. Cleveland, OH: Freshwater Press.
Donahue, James (1995) *Steamboats in Ice 1872*. Cass City, MI: Anchor Publications.
    (1990) *Steaming Through Smoke and Fire 1871*. Ann Arbor, MI: Historical Society of Michigan.
Gentile, Gary (1995) *The Nautical Cyclopedia*. Philadelphia, PA: Gary Gentile Productions.
Green, J B (1859) *Diving With and Without Armor*. Buffalo, NY: Steam Power Press.
Greenwood, John O (1987) *Namesakes 1900–1909*. Cleveland, OH: Freshwater Press.
    (1986) *Namesakes 1910–1919*. Cleveland, OH: Freshwater Press.
    (1984) *Namesakes 1920–1929*. Cleveland, OH: Freshwater Press.
    (1995) *Namesakes 1930–1955*. Cleveland, OH: Freshwater Press.
Hall, Captain J W (1872) *Maritime Disasters of the Western Lakes*. Detroit, MI: Free Press Book and Job Printing Establishment.
Kadar, Wayne Louis (2005) *Great Lakes Passenger Ship Disasters*. Gwinn, MI: Avery Color Studios.
Kohl, Cris (1998) *100 Best Great Lakes Shipwrecks, Volume I*. West Chicago, IL: Seawolf Communications.
    (1990) *Dive Ontario*. Chatham, ON, CA: Cris Kohl.
    (1988) *Dive Southwestern Ontario*. Chatham, ON, CA: Cris Kohl.
    (2004) *Shipwreck Tales of the Great Lakes*. West Chicago, IL: Seawolf Communications.
    (2008) *Great Lakes Diving Guide. Enlarged 2nd Edition*. West Chicago, IL: Seawolf Communications.
Kohl, Cris and Joan Forsberg (2016) *Great Lakes Shipwrecks*. West Chicago, IL: Seawolf Communications.
Kuntz, Jerry (2016) *The Heroic Age of Diving: America's Underwater Pioneers and the Great Wrecks of Lake Erie*. Albany, NY: State University of New York Press.

Lytle, William (1952) *Merchant Steam Vessels of the United States 1807–1868*. Mystic, CT: Steamship Historical Society of America.

Mansfield, John Brandt (1899) *History of the Great Lakes, Volume I*. Chicago, IL: J H Beers & Co.

(1899) *History of the Great Lakes, Volume II*. Chicago, IL: J H Beers & Co.

Oickle, Alvin F (2011) *Disaster on Lake Erie: The 1841 Wreck of the Steamship Erie*. Charleston, SC: The History Press.

Petkovic, Erik A (2017) *Shipwrecks of Lake Erie Volume One*. Huntingtown, MD: Fathom Productions.

Polk, R L (1888) *R L Polk & Company's Marine Directory of the Great Lakes*. Detroit, MI: R L Polk & Company.

Ratigan, William (1977) *Great Lakes Shipwrecks & Survivals*. Grand Rapids, MI: William B Eerdmans Publishing Co.

Stanton, Samuel Ward (1895) *American Steam Vessels*. New York, NY: Smith & Stanton.

Stonehouse, Frederick (1984) *Went Missing: Unsolved Great Lakes Shipwreck Mysteries*. Marquette, MI: Avery Color Studios.

Wachter, Georgann and Michael (2003) *Erie Wrecks East*. 2nd Edition. Avon Lake, OH: Corporate Impact.

(2007) *Erie Wrecks & Lights*. Avon Lake, OH: Corporate Impact.

Newspapers

Akron Beacon Journal

Amherstburg Echo

Buffalo Commercial Advertiser

Buffalo Courier

Buffalo Daily Courier

Buffalo Daily Republic

Buffalo Enquirer

Buffalo Evening News

Buffalo Express

Buffalo Gazette & Niagara Intelligencer

Buffalo Morning Express

Chicago Inter-Ocean

Chicago Tribune

Cleveland Herald & Gazette

Cleveland Morning Leader

Cleveland Plain Dealer

Cleveland Press

Daily British Whig

Daily National Pilot
Detroit Advertiser & Tribune
Detroit Free Press
Detroit Gazette
Detroit Post
Detroit Post & Tribune
Duluth News Tribune
Erie Daily Dispatch
Erie Observer
Erie Weekly Gazette
Great Lakes News
Lorain Morning Journal
Milwaukee Evening Wisconsin
Milwaukee Daily Sentinel
New York Times
Port Colborne News
Port Huron Times
Saint Catherine's Evening Journal
Saginaw Courier Herald
Sheboygan Press
Titusville Herald
Toledo Blade
Toronto Globe
Weekly Buffalonian

Periodicals and journals

Board of Lake Underwriters
Board of Marine Inspectors
Canadian Law Times
Federal Reporter
Green's Marine Directory of the Great Lakes
Inland Lloyds Register
Inland Seas
Inland Waterways and Seaway Journal
Lytle Holdcamper List
Marine Record
Marine Review
Merchant Vessel Lists

Wisconsin Marine Historical Society Soundings
World Wide Magazine
Wreck Diving Magazine

Other sources

Becky Schott
Cleveland State University
Cris Kohl
Department of Marine
Gary Gentile
Garry Kozak
Library of Congress
Milwaukee Public Library
National Archives
National Archives at Chicago
National Archives at Philadelphia
Pat Clyne
Peter Hess
Rutherford B Hayes Presidential Library
Steve Gatto
Tom Wilson
United States Circuit Court of Appeals, Second District
United States Circuit Court of Appeals, Sixth District
United States Coast Guard Casualty Lists
United States Coast Guard District Nine
United States Coast Guard Historian's Office
United States District Court, Eastern District of Michigan
United States District Court, Northern District of Ohio
United States District Court, Western District of New York
United States Steamboat Inspection Service
United States Supreme Court
Vlada Dekina
Warren Lo
Wisconsin Marine Historical Society

Contributors

Photographers

Please support the following contributing photographers whose spectacular underwater images greatly helped to enhance this book. Visit their websites, purchase their books, read their articles, retain their services, buy their photos, take their classes and join their trips.

Gary Gentile — www.ggentile.com

Cris Kohl — www.seawolfcommunications.com

Becky Schott — www.liquidproductions.com and www.megdiver.com

Paul Lenharr — www.somddivers.com and www.lenharr.com

Warren Lo — www.warrenlophotography.com

Tom Wilson — www.scubaq.ca

Steve Gatto — *Where Divers Dare: The Hunt For The Last U-boat* (U-550). Author of the forthcoming *Tugboat Down*.

Vlada Dekina — www.wrecksandreefs.com

Side scan sonar images

Garry Kozak — www.2kozak.com

Index

A

A A Turner Shipyard *150–151*
Abandoned Shipwreck Act 1987 *61*, *63*
Abernathy, Andrew *127*
abolitionist movement *66*
Acme *21–27*
Admiral *17*
Admiralty Wreck *70*
advertisements *33*, *148*
A-frames *29*, *82*
Airport, Ohio *110*
Albert J Wright *143–146*
Alleghany *135*
Amasa Stone *83–84*
American Express Company *42*, *50–51*
American Transportation Company *110*
Amos, Art *124*
anchors *54*, *56*, *59*, *68*, *71*, *107*, *112*, *129*, *142*, *147*, *156*
Andrew B *28–29*
Anna Smith *120*
Argo Steamship Company *73*, *77*
Arrow *105*
artifacts *61–62*, *69*, *88*, *112*, *130*, *142*
Ashtabula, Ohio *24*
Atlanta *25*
Atlantic *26*, *30–51*, *105*

B

Bagnall & Dobbins *125–126*
Barcelona, New York *25*, *92*, *93*
Barge F *52–54*
barges *52–54*, *82*, *120*
barks *12*, *25*, *131–142*
Barry, Morris *34*

Battle of Lake Erie *65–66*
Bayersher *81*
Bay State *109*
Bay Steamship Company (UK) *81*
beachings *126*, *135*
Beeson's Marine Directory for the Northwestern Lakes *148*
belaying pins *12*, *112*, *113*, *122*
bells *35*, *111*, *142*
bends. See *decompression sickness*
Benjamin B Jones shipbuilders *104*
Bennett *135*
Berryman, John *125–126*
Betty & Jean *93*
Birmingham, UK *32*
Bishop, Albert *44*
Bissennette, Joseph *92*
bitt *12*, *13*
Black Rock *25*
Blake *135–136*
Blechele, Carl A *87*
"Bob Powell's Wreck" *102*
boilers *32*, *74*, *107*, *116*, *121*, *146*, *156*
Bois Blanc Island, Michigan *135*
booms *68*
border *30*
Bournes Beach, New York *93*
bowsprits *12*, *56*, *67*, *68*, *69*, *100*, *102*, *109*, *122*, *129*, *142*
bow stems *12*, *121*
braces *12*
breaking records *117*
brigs *25*, *65*, *108–114*
British armed forces *65*, *131*
British North West Trading Company *65*

Brooks, Harry *87*
Brooks, Joseph Sidney *93*, *96*
Brooks, Lewis *96–97*
Brown, Charles *111*
Bruce Mines, Ontario *28*
Buffalo Commercial Advertiser *26*, *32*, *63*, *116–117*, *152*
Buffalo Courier *120*
Buffalo Daily Courier *22*, *23*, *25*, *32*
Buffalo Daily Republic *38*, *39–40*, *45*, *47–48*, *124*
Buffalo Enquirer *153*
Buffalo Evening News *74*
Buffalo Express *104*, *128*
Buffalo Gazette & Niagara Intelligencer *66*
Buffalo Morning Express *120*, *144*, *145*
Buffalo Mutual Insurance Company *125*
Buffalo, New York *21*, *25*, *26*, *32*, *33*, *34*, *64*, *65*, *66*, *67*, *74*, *104*, *137*, *139*, *144*
Buffalo Steamship Company *76*
Buffalo & Susquehanna Coal and Iron Company *77*
Burrill, Captain John *126*
Butler *152*
Byers, William *92*

C

cabins *55*, *56*, *69*, *70*, *71*, *95*, *123*, *128*
Cadet *104*
Caledonia *61*, *65–66*
Calkins, Almon *37–38*
Calumet *146*
Canadian Dredge and Dock Company *28*
canalers *12*, *73–74*, *81*, *90*
capstans *12*, *53*, *123*, *130*, *156*
captains (authority of) *32–33*, *94*, *153*

cargo *24*, *26*, *32–34*, *61*, *62*, *64*, *65*, *66*, *73*, *76*, *82*, *86*, *91*, *92–94*, *105*, *109*, *110*, *116*, *117*, *118*, *124*, *126*, *134*, *150*, *152*, *153*, *154*, *155*
Carney, James *34–35*
Casey, Thomas *117*
Castle *135*
Cataract *110–112*
centerboards *113*, *158*
chains *68*
Charles Donnelly *84*
Chaumont, New York *108*
Cheboygan, Michigan *135*
Chesapeake Bay *93*
Chicago, Illinois *23*, *24*, *66*, *104*, *117*, *126*, *134*, *146*, *152*, *153*
Chicago Inter Ocean *135*
City of Charlevoix *146*
City of Erie *137–139*
Civil War *23*, *50*
Claremont *81–82*
Cleveco *17*
Cleveland and Buffalo Transit Company *139*, *140*
Cleveland Herald *124*, *144*, *152*
Cleveland, Ohio *23*, *34*, *64*, *73*, *104*, *105*, *110*, *115*, *120*, *137*
Clyne, Pat *60*, *61*
Coady, Quayle & Company *115*
cofferdam *12*, *78*
Colberg, Peter *23*
collisions *35*, *37*, *76*, *79*, *110–112*, *128*, *135–136*, *138–140*, *157*
companionways *53*, *56*, *120*, *147*
compass *62*
Connolly, Madeline *138*
controversy *30*
Copley and Mann shipyard *108*
Cormorant *120*

Corning 25

coroner 25

Cortland 19

court cases. See *legal battles*

Cracker 55–57

crane 29

crew 86, 87, 92, 105, 106, 117, 118, 119, 127, 138, 154, 155

　not doing duty 36–37

Cromwell, Rob 120

crosstrees 12, 68, 113

"Crow's Nest" wreck 113

cruising 145

Cunning, Captain Alex 77, 78

currents 22, 102, 121

cutwater bow 12, 112

D

Dacotah 153, 154

Daily British Whig 135, 136

Daly, Nicholas 135

damage/cost 23, 24, 35, 36, 80, 82, 105, 109, 111, 120, 125, 126, 135–136, 146, 152, 154

davits 67, 95, 147

Dayton 63–64

deadeyes 12, 68, 112, 113, 122, 130, 141

Dean Richmond 52, 116

decompression sickness 43, 48–49

Dekina, Vlada 52

Demers, L A 94

Department of Marine (Canada) 94

derricks 135

　Bishop derrick 44

Detroit Advertiser 44–45

Detroit Free Press 39, 151

Detroit Gazette 66

Detroit, Michigan 32, 65, 124, 150, 154, 155

Detroit Post 25–26

Detroit River 23, 24, 65, 75–76

Detroit Tribune 146

Dickson, Captain William 24–25

Dickson, John 66

Dickson Tavern 66

diving bell 42

diving equipment 42, 47–48

　helmet 58

Dominion Sugar Company 81

Dominion Wreck Commissioner 94

Donner, Ed 87

Douglass, Mr 137

Doyle, Captain James 145

dragging 49

dredges 62

　barge 28–29

drunk sailors 136

Duluth News Tribune 155

DuMurs, Ned 87

Dunkirk, New York 59, 84, 106

Dunkirk Schooner 58–72

E

East Sister Island 109

Easy, A M 119

Emerson, Michigan 155

Emling, M J 87

Empire State 118–120

Encorse, Michigan 78

Endleman, Robert 83

engine rooms 95

engines 27, 32, 74, 107, 116, 118, 121, 146, 147, 156

English Channel 81

Erie 26

Erie, Pennsylvania 64, 91, 106, 124

Erwin L Fisher 73–81

Evans, John 117

F

Fairport Harbor, Ohio *105*, *137*
Falkonet, Jack *22*
fantails *12*, *156*
fatalities *23*, *31*, *36*, *64*, *65*, *83*, *84*, *85*, *93*, *106*, *111*, *137–140*, *138*
Favorite *77*
fife rails *12*, *69*, *130*, *142*
figureheads *12*, *56*, *60*, *68*, *69*, *122*, *129*
fire *117–119*, *121*, *128*, *146*, *151*, *156*
Fisher, Erwin L *73*
Fletcher, Mike *50*
Flint, Captain Samuel *117–119*
flotsam *26*
fog *34*, *92*, *152*
Follett, Captain Edward *64*
Fort Dearborn *65*
Fort Erie *65*
Fortier, Labon B *143*, *145*
Fort Michilimackinac *65*
Fred Kelley *152*
freighters *90*, *104–107*, *116–121*

G

galley *95*
Gallinipper *105*
Garner, William Thomas *138*, *140*, *142*
gas well *100*
Gatto, Steve *58*, *68*, *69*, *70*
Geary, George *93*
General Harrison *64–65*
General Wayne *66*
General Worth *25*
Gentile, Gary *62*, *71*, *72*
George J Whelan *73–89*
George W Bissell *23*
George Worthington *25*
Gettysburg Adams Sentinel *41*

Gilchrist, Ralph *81*
Gleason's Pictorial *37*
Globe Iron Works *116*
Godfrey, Charles *87*
Godfrey, Mrs Charles *87*
Grace Murray *135*
Grand Island, New York *63*
Grand Traverse Bay *22*
Grape Shot *23*
Grassy Island *76*
Gray & Gray *154*
Greater Detroit *84*
Great Lakes Engine Works Yard *78*
Great Lakes Towing Company *77*
Green, John B *42–43*, *47–50*
Griffon *60*
Grosse Ile Channel *76*
groundings *81–82*
Gulnare *136*
gunwales/gunnels *12*, *54*, *128*, *141*

H

Halstead *135*
Hamilton *60*, *135–136*
Hardison, George *21*
Harrington, Elliott *49–50*
hatches *68*, *69*, *82*, *113*, *118*, *129*, *156*
Hawman, Captain Edward C *92*, *94*
hawsepipes *12*, *56*, *147*
helm *52*, *143*, *147*. See also *wheels*
Henry, Joseph W *53*
Herbert, Captain Jim *52*, *88*
Hess, Peter *58–59*
Hogg & Delameter *32*
hogging arches *12*, *21*, *27*, *51*, *104*, *107*
holds *53*, *56*, *82*, *87*, *95*, *107*, *116*
Hollis, Thomas *138*
Holy Bible *62*

horns 79
Hulgate, Samuel 106
human remains 61, 62
Humble, John 143–144
Hunshedt, Boatswain Sivert E 93
Hunt, Captain 111

I

Idaho 107, 136–137
immigrants 33–34
I N Foster 146
Innes, Duncan & Cowan 153
Interlake Steamship Lines 84
Iron City 125
Iron Fish 101

J

jewelry 62
John J Boland Jr 90–99
John Wolverton 31
Jones, Harry 93
Junction 20 100–101

K

Kane, John 19
keel 67
Kelley Island Lime and Transportation Company 82, 87
Kelleys Island 105
Keystone State 105–106
Kingston News 106
Kingston, Ontario 129, 134
Kohl, Cris 113, 114
Kozak, Garry 52, 88–89, 115, 121, 122, 130, 140, 141
Kullberg, Richard 59–63

L

Lake Erie Boiler Works 74
Lake Huron 28, 75, 153
Lake Michigan 23, 152
Lake Ontario 82, 91, 108, 110, 111, 115, 126, 135, 136, 152
Lake Saint Clair 75
Lake Shore Elevators 152
Lake Superior 136, 155
Lake Transportation Company 81
Lama, Tony 157
Lange, Eckhart 83
La Salle, Robert de 60
leaks 86, 128
Leatham & Smith Towing & Wrecking Company 146
Lee, Captain John 110–111
legal battles 30, 41–42, 50, 60, 75, 78–80, 140
Lennae, William E 155
lifeboats 13, 36–37, 38, 64, 70, 83, 86–87, 89, 92, 94, 106, 118, 128, 131, 137, 138, 141, 155
life preservers 39
   Ward Life Preserver (wooden stool) 40–41
lighters 77
lighthouses 100
lights (ships) 23, 34, 38, 110, 111–112, 138–140
listing 83, 86
Liverpool, UK 32
Lloyds Register 90
loading 32–34, 91
Longnecker, W P 87
Long Point, Ontario 100–101, 102, 110, 129, 155
   Lighthouse 34, 119
Looker, Frank 82
Lorain, Ohio 76
Love, James 118

Lo, Warren *73*, *89*
lumber hauler *150*
Lyme, New York *65*

M

MacIntyre, Jean *93*
Mackinac, Michigan *65*
Magee, Kevin *52*
Maillefert, Benjamin *42*
Malta *131*, *132*
manifests *34*
Manistique *146*
Manitou *153*, *155–156*
Manitoulin Island, Ontario *153*
Mansfield *110*
Marine Cigar *44–46*
Marine City, Michigan *31*
Mariposa *137*
Marquette & Bessemer No 2 *121*
Marquette, Michigan *124*
Mast Hoop *102*
mast hoops *102*, *122*
masts *12*, *13*, *51*, *57*, *68*, *69*, *72*, *100*, *101*, *102*, *107*, *109*, *112*, *116*, *122*, *130*, *141–142*, *146*, *149*, *150*, *156*
Mayflower *105*
McAlpine, Captain *137–138*
McCallum, James *138*
McKenna, Captain *120*
McNeill, Captain Walter H *83–84*, *86*
McNell, DeGrass *34–35*
measuring (a ship) *67*
Merrick *118*
Merry & Gay Shipyard *124*
Michigan Central Dock *45*
Michigan Central Railroad Company *31*, *34*
Milan, Ohio *124*
Milligan, Captain Alex *136*

Mills & Co *144*
Milwaukee, Wisconsin *22–23*, *117*
Misener, Captain *81*
Montana *118*
Montreal, Quebec *53*, *94*, *146*
Mountaineer *25*
Muskegon, Michigan *152*, *153*

N

Nash, Jacob *37*
National Park Service *63*
National Register of Historic Places *63*
Nautilus *42*
Nebish Rapids *124*
Neucheler, William *87*
New Orleans *104*
New York *24*
New York, State of *61*, *62*, *63*, *66*
Niagara River *65*, *145*
    Niagara Falls *91*
Nolan, Captain William P *86*, *87*
Northeast Research LLC (NER) *59–63*, *67*
Northerner *111*
Northern Transportation Company *34*

O

Ocean *45*
Offshore Supplier *28*
Ogdensburg *34–38*
Ohio *105*, *107*
Ohlemacher, Irving *83–84*, *85–86*
Oneida *103–107*
Ontario *110*
Ontario, Canada *65*
Orient Insurance Company *120*
Osprey Charters *10*
Oswego, New York *34*, *111*, *126*, *136*
Oswego Times and Journal *109*, *111*

Oxford 108–114

P

paddle wheels *51*
Panic of 1873 *151*
Parsons *135*
Pashaw, Clarence *76–77*
passengers *32–34, 36–37, 45, 117, 118, 119*
   deck *32*
   first class *34*
   steerage *32, 34*
Pearce, Thomas *87*
Pease & Allen & Burke *104*
Pelee Island *109*
penetration *18, 89, 95, 107, 144, 147, 151, 156*
Pennsylvania *65*
People's Line *22*
Perry, Commodore *65*
Persian 115–121
Pettey, Captain J Byron *34, 36–37*
Phillips, Lodner G *44–46*
Pickel, Edward S *137*
picket boats *84*
Pierce, Thomas *85*
Pigeon Island *135*
Pilot Rules (of the Great Lakes) *78*
Pittsburgh & Erie Coal Company *140*
pocket watch *61*
Point Abino, Ontario *65*
Point Pelee, Ontario *22, 24, 120*
Point Prophey, Lake Superior *136*
Port Colborne, Ontario *25, 138, 140, 146*
Port Dalhousie, Ontario *136*
Port De Caen *81*
Port Dover, Ontario *48*
portholes *17, 88, 95*
Port Stanley, Ontario *28, 152*
Port Weller Dry Dock *28*

Presque Isle, Pennsylvania *66*
propeller *13, 21–27, 88, 95, 107, 121*
P S Marsh *25*
pumps *53, 54, 56, 69, 71, 113, 118, 123, 125–126*

Q

Quayle & Martin Shipbuilders *115*
Quayle & Moses *115*
Quayle & Sons *116*
Quayle, Thomas *115*
Queen City *25*
Quinlan, Captain W P *153, 154, 155*
Quinn, Captain John *154*

R

Racine, Wisconsin *64*
railings *68, 113, 156, 158*
raising wrecks *77*
ratings *9*
Reed, General Charles M *124, 127*
Reed Mansion *66*
Reed, Rufus *63, 66*
Register of Historic Places (New York) *63*
regulations *32, 94, 153*
Rescue *26*
research *18–19, 21, 63, 103–104*
Richards, Mr (coroner) *26–27*
Richardson *136*
Richardson, George *125*
Rich, Captain William *106*
Richmond, Dean *122*
rigging *112, 120*
River Rouge *65*
Robert B *85*
Rogers, H R *139*
Root, Captain Frank D *137–138*
Roscoe *64*
Roth *77*

R R Johnson 110
rudder posts *27, 55, 70, 95, 121, 156*
rudders *56, 70, 71, 89, 95, 113, 121, 123, 130, 141*

S

Sadler, Brownie *93*
safe *26, 42, 47–48*
Saginaw Courier *153*
Saint Catherines Evening Journal *132*
Saint Catherines, Ontario *90, 132*
Saint Clair, Michigan *64*
Saint Clair River *75, 153*
Saint James *122–130*
Saint Lawrence River *82, 91*
Saint Mary's River *124*
salvage *30, 42–50, 60–63, 77–78, 80, 146, 154*
  divers *42–44, 47–50, 61, 157*
samson posts *13, 54, 68*
Samuel and Eber Ward *31, 41–42*. See also *life preservers: Ward Life Preserver (wooden stool)*; and *Ward, Eber*
sand-suckers *73–89*
Sandusky, Ohio *82*
Sandusky Register *86*
Sandusky Star Journal *85, 87*
Sarah E Bryant *120*
Sarnia, Ontario *146*
Sarnia Steamships *90*
Sault Ste Marie, Ontario *155*
schooner barges *13*
Schooner G. See *Dunkirk Schooner*
schooners *13, 25, 52–54, 58–72, 102, 104–105, 118, 122–130, 149, 157–158*. See also *schooner barges*
Schrum, Ross *157*
Scourge *60*
scows *13, 55–57, 100–101*

scroll heads *12, 64, 65, 122*
scuttling *127*
Sears, Henry *42*
shallowest lake *29, 92*
Shickluna, Louis *132*
Shickluna Shipyard *132–134*
side scan sonar *45, 88–89, 115, 121, 122, 130, 141, 157*
sightseeing *145*
signals *153*
Sincennes McNaughton Company (Sinmac) *146*
Sir C T van Straubenzee *131–142*
Sir William Siemens *154*
slaves *66*
Smith *143–148*
Smith, M *92*
smokestacks *17, 43, 147, 150, 156*
South America *65*
Southwind *10*
spars *13, 110*
speed *32, 129*
SS *Norwood* *93*
SS *Texana* *93*
stairs *53, 147, 156*
Stamm, Arthur *83*
Stanley, John *87*
staves *64*
steamships *21–27, 30–51, 90–99, 104–107, 109, 116–121, 137–139, 143–148, 152, 154*
steam whistle *147, 153*
steering *52, 141*
  chain *52*
  drum *52*
Stephen M Clement *76*
Stern Castle *149*
stern gallery windows *60*
St Lawrence Foundry *53*

St Louis 105
Stone, Captain James 139–140
storms 17, 22, 23, 28, 44, 64, 65, 82–83, 86, 92, 100, 105, 109, 110, 128–129, 137, 146, 152, 153, 155
    November gales 22, 24–25, 64
Straits of Mackinaw 126
strandings 151
Sturgeon Point 25
submarines 44–46, 120
Sudgen, Kate 76, 78
Sudgen, Lewis 76, 78
superstructure 95, 147, 150
Supply 25
Swallow 150–156
Swan, Hunter and Wigham Richardson Ltd (Swan Hunter) 90, 96

T

T-8 157–158
"taking the Fifth" 135
Talcott & Underhill 123
Talmage, F 136
Tashmoo 137
Texas 13, 119
Thomas Kingsford 126
Thompson, C D 146
tiercins 13, 105
tillers 13, 69, 113, 141
Titanic 94
Titusville Herald 126, 129–130
Todd, Captain James M 86, 87–88
Toledo Blade 127
Toledo, Ohio 126, 151
Toledo Shipbuilding Company 73, 80
Toronto Globe 119, 134
Toronto, Ontario 28
towing 119, 136, 146, 155

Townsend, Captain Henry H 76, 77, 78
transoms 13, 65, 70, 71, 100, 113, 130, 147
treasure 47, 49, 62
Trenton, Michigan 150
Truman, Thomas 111
tugs 17, 28, 77, 118, 120, 135, 136, 146
turnbuckles 13, 54
turtle (turned) 13, 83, 93, 105, 106
Tuttle, Judge Arthur J 78
Tyneville 90

U

Underground Railroad 66
underwater photography 52, 68, 70, 89, 112, 129, 141, 142, 147, 149
Union Dry Dock Company 143
United States Coast Guard (USCG) 161 93
    Cleveland 84
    Erie 84
United States Steamboat Inspection Service 76, 78, 87, 111, 139, 154
US Signal Office 128

V

van Straubenzee, Charles Thomas 131
VanZandt, David 52
Vermillion, Ohio 65
visibility 29, 51, 54, 68, 88, 95, 121, 147, 156

W

Waage, Captain Thomas 82–83, 86–87
walking beam 13
Walters, Arthur 83, 84, 87
Walter Scranton 155
Ward, Eber 40. See also *Samuel and Eber Ward*
War of 1812 60, 65, 115

Warrington, Judge *78–79*

Weis, Ralph *87*

Welland Canal *34*, *91*, *126*

Wells, Henry *42–43*

Western Transit Corporation *137*

wheelhouses *17*, *143*, *147*

wheels *52*, *130*, *141*, *143*, *149*

White *77*

white eagle *65*

White Law *154*

Whitney, D Jr *150*

William, C *106*

William John *128*

Wilson, Tom *55*, *122*, *131*, *143*

winches *54*, *95*, *113*, *129*

windlasses *13*, *27*, *54*, *56*, *68*, *107*, *142*, *147*, *149*

windows *70*, *71*

Windsor, Ontario *65*

World Wars *17*, *73*, *81*

wreck diving *17–19*. See also *salvage: divers*
 wreck diver code *50*

Wright, Captain *119*

W W Arnold *25*

Y

yawl boats. See *lifeboats*

Z

zebra mussels *29*, *51*, *157*

Zeck, Arthur *87*

Jonas Arvidsson

DIVING EQUIPMENT
CHOICE, MAINTENANCE AND FUNCTION
Paperback and ebook

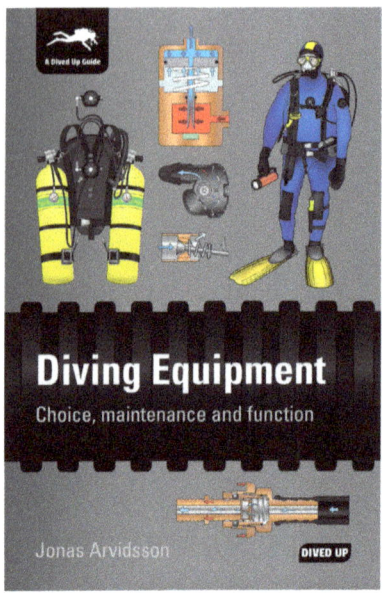

Diving Equipment is a guide to how diving gear works. It is completely independent of brands — focusing instead on the design, technology and practicality of the equipment options available for scuba and other types of diving. Former Head of Education at PADI Sweden Jonas Arvidsson details the choices that divers are faced with and gives tips to help ensure your gear will last. If you want to better understand diving kit without having to dismantle it, then *Diving Equipment* — with its colourful technical illustrations and explanations of the hidden inner workings — is the answer.

ISBN 978-1-909455-13-9 | 2016 | 2nd Edition

Nick Lyon

THE DIVER'S TALE
Paperback and ebook

Britain is an island nation so, unsurprisingly, scuba diving is a popular British pastime enjoyed by some 50,000 keen participants and just as many of the armchair variety. A carefully structured program of training ensures that the nation's divers are well prepared for the challenging conditions which may be encountered beneath our seas. Or does it? How many trainee divers were taught about the perils of high-speed testicular trauma during descent? Or of having sex in a tent with a deaf person? Why bacon should be in your first aid kit? How to build a space shuttle using salvaged ammunition? *The Diver's Tale* is the unvarnished account of real British diving with humorous, embarrassing and tragic true tales based on the author's 40 plus years of experience.

ISBN 978-1-909455-24-5 | 2019 | 2nd Edition

www.DivedUp.com

www.ingramcontent.com/pod-product-compliance
Lightning Source LLC
Chambersburg PA
CBHW041952180426
43199CB00038B/2890